T0280701

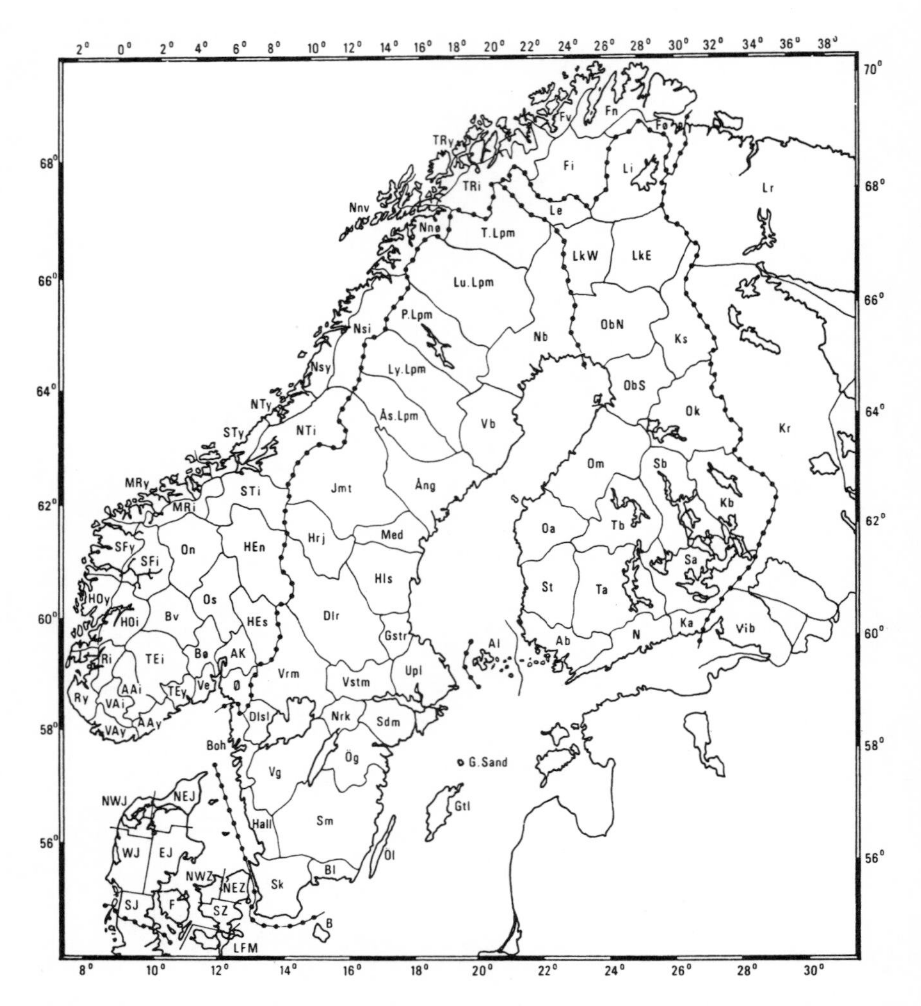

List of abbreviations for the provinces used throughout the text, on the map and in the following tables.

DENMARK

SJ	South Jutland	LFM	Lolland, Falster, Møn
EJ	East Jutland	SZ	South Zealand
WJ	West Jutland	NWZ	North West Zealand
NWJ	North West Jutland	NEZ	North East Zealand
NEJ	North East Jutland	B	Bornholm
F	Funen		

SWEDEN

Sk.	Skåne	Vrm.	Värmland
Bl.	Blekinge	Dlr.	Dalarna
Hall.	Halland	Gstr.	Gästrikland
Sm.	Småland	Hls.	Hälsingland
Öl.	Öland	Med.	Medelpad
Gtl.	Gotland	Hrj.	Härjedalen
G. Sand.	Gotska Sandön	Jmt.	Jämtland
Ög.	Östergötland	Ång.	Ångermanland
Vg.	Västergötland	Vb.	Västerbotten
Boh.	Bohuslän	Nb.	Norrbotten
Dlsl.	Dalsland	Ås. Lpm.	Åsele Lappmark
Nrk.	Närke	Ly. Lpm.	Lycksele Lappmark
Sdm.	Södermanland	P. Lpm.	Pite Lappmark
Upl.	Uppland	Lu. Lpm.	Lule Lappmark
Vstm.	Västmanland	T. Lpm.	Torne Lappmark

NORWAY

Ø	Østfold	HO	Hordaland
AK	Akershus	SF	Sogn og Fjordane
HE	Hedmark	MR	Møre og Romsdal
O	Opland	ST	Sør-Trøndelag
B	Buskerud	NT	Nord-Trøndelag
VE	Vestfold	Ns	southern Nordland
TE	Telemark	Nn	northern Nordland
AA	Aust-Agder	TR	Troms
VA	Vest-Agder	F	Finnmark
R	Rogaland		

n northern   s southern   ø eastern   v western   y outer   i inner

FINLAND

Al	Alandia	Kb	Karelia borealis
Ab	Regio aboensis	Om	Ostrobottnia media
N	Nylandia	Ok	Ostrobottnia kajanensis
Ka	Karelia australis	ObS	Ostrobottnia borealis, S part
St	Satakunta	ObN	Ostrobottnia borealis, N part
Ta	Tavastia australis	Ks	Kuusamo
Sa	Savonia australis	LkW	Lapponia kemensis, W part
Oa	Ostrobottnia australis	LkE	Lapponia kemensis, E part
Tb	Tavastia borealis	Li	Lapponia inarensis
Sb	Savonia borealis	Le	Lapponia enontekiensis

USSR

Vib	Regio Viburgensis	Kr	Karelia rossica	Lr	Lapponia rossica

FAUNA ENTOMOLOGICA SCANDINAVICA
Volume 9 1980

The Aphidoidea (Hemiptera) of Fennoscandia and Denmark. I

General Part.
The Families Mindaridae, Hormaphididae, Thelaxidae, Anoeciidae, and Pemphigidae

by

Ole E. Heie

SCANDINAVIAN SCIENCE PRESS LTD.
Klampenborg . Denmark

Fauna entomologica scandinavica
is edited by »Societas entomologica scandinavica«

World list abbreviation
Fauna ent. scand.

Printed by
Vinderup Bogtrykkeri A/S
7830 Vinderup, Denmark

ISBN 87-87491-34-6
ISSN 0106-8377

Contents

Two colour plates are given on p. 99 and p. 101.

Preface

Volume 9 of "Fauna entomologica scandinavica" is the first of four planned volumes dealing with the superfamily Aphidoidea of the infraorder Aphidodea (= Aphidoidea of Börner) within the order Homoptera, suborder Sternorrhyncha.

The volume comprises the introductory chapters to the superfamily and the systematic treatment of the families Mindaridae, Hormaphididae, Thelaxidae, Anoeciidae, and Pemphigidae. The next volume on Aphidoidea deals with the Drepanosiphidae (including the subfamilies Drepanosiphinae, Phyllaphidinae, and Chaitophorinae). The third volume deals with part of the Aphididae, viz. Pterocommatinae and Aphidinae: Aphidini. The fourth volume may appear in two parts, dealing with the remaining species, viz. Aphidinae: Macrosiphini and Lachnidae.

A zoogeographical catalogue is given in each of the four volumes, references are given in each volume, but those listed in the first, second, and third volumes do not reappear in the second, third, and fourth volumes, respectively. The last volume contains a host plant index of all species treated in the four volumes. Each volume has its own index of entomological names. The last volume is concluded with an index listing all entomological names in the four volumes.

The Scandinavian species are provided with numbers continuing throughout the four volumes, but each volume is paged separately.

The following geographical terminology is used:

Scandinavian Peninsula: Norway and Sweden
Fennoscandia: Scandinavian Peninsula, Finland, and adjacent parts of the USSR
Scandinavia: Fennoscandia and Denmark
Nordic countries: Scandinavia, Iceland, and the Faroes
Spitzbergen is not included in Norway
Greenland and the Faroes are not included in Denmark
N Germany: West Germany north of Hamburg and East Germany north of 53°N latitude.

The following abbreviations are used in the descriptions:

ant. segm.: antennal segment
abd. segm.: abdominal segment
IIIbd.: basal diameter of ant. segm. III (Fig. 57b)
VIa: basal part of ant. segm. VI (from base to distal margin of primary rhinarium) (Fig. 5)
VIb (processus terminalis): the ultimate thinner part of ant. segm. VI
2sht.: second segment of hind tarsus.

Abbreviations in figures only:

apt. viv.: apterous viviparous (and parthenogenetic) female
al. viv.: alate viviparous (and parthenogenetic) female
ovip.: oviparous (not parthenogenetic) female

Botanical names are mostly in accordance with Lid (1963). Synonyms used by entomological authors may be added in brackets, especially if used in original descriptions of aphid species. Authors (or generally accepted abbreviations of authors) of botanical names are given in the host plant index. Host plant names are given in the descriptions and host plant index, even if the species in question has not been found on this plant or on these plants in Scandinavia, unless 1) the host plants of a species are unknown, 2) if they do not occur in Scandinavia (or have not been introduced), or 3) if the number of host plants is great (in polyphagous species and some oligophagous species); in the latter case the host plant genera — or some of them — are given. Some host plant records are left out, because they seem to be exceptional, maybe due to misidentification, or having been referred to observations on accidental landing of alate aphids.

Keys to all families, subfamilies, and tribes occurring in Scandinavia are given in this volume. The description of each family is followed by a key to subfamilies and occasionally to tribes. Additional keys to tribes are presented in some cases under the respective subfamilies, or are left out (if the Scandinavian species of a subfamily belong to the same tribe). Keys to genera are given in connection with descriptions of the higher taxa to which they belong, and keys to species are given in connection with descriptions of the genera to which they belong.

Keys are provided for apterous as well as for alate viviparous females, unless one of these morphs is little known or does not exist (the alate female is the only viviparous morph in some genera), but not for sexuales, which are still unknown in many species.

Identification of single specimens by keys is difficult, sometimes impossible. When possible, several specimens belonging to the same colony, should be examined. Even then identification may be unsafe, if the host plant is unknown. The range of intraspecific variation of morphological characters is often considerable; specimens belonging to the same morph of some mutually related species are very similar even if the species are ecologically different.

The text is based not only on my own studies of specimens but also on litterature so that the range of variation within each morph of each species appears from the description. Information on the biology has been compiled from the literature and own observations. The papers from which information has been taken are numerous; they are mentioned in the references. They include original descriptions of taxa, revisions, monographs, etc., cf. Börner (1930, 1952), Börner & Heinze (1957), Doncaster (1961, 1973), Eastop (1961, 1966), Hille Ris Lambers (1938, 1939a, 1939b, 1947a, 1947b, 1949, 1953), Mordvilko (1935), Palmer (1952), Stroyan (1950, 1955, 1957, 1964a, 1964c, 1966, 1972, 1977), Tullgren (1909, 1925), and Zwölfer (1957–1958).

Synonyms in older Scandinavian literature are given. Additional synonyms may be found in Eastop & Hille Ris Lambers (1976), in the descriptions referred to as the "Survey".

The treatment of each species usually consists of a description of the apterous viviparous female and additional descriptions of other morphs. The term "apterous viviparous female" in keys and descriptions does not cover the fundatrix unless definitely expressed.

For each species the distribution in and outside Scandinavia is given on the basis of the literature and supplementary information not previously published. The published records have mainly been found in the following papers: Finland: Heikinheimo (1944, 1963, 1966a), Thuneberg (1960, 1962, 1963), Heie & Heikinheimo (1966); Norway: Tambs-Lyche (1968, 1970 a.o.), Ossiannilsson (1962), Heikinheimo (1966b); Sweden: Ossiannilsson (1959, 1969b); Denmark: Heie (1960-70, 1972a, 1973a, 1976b). — Iceland: Hille Ris Lambers (1955), Prior & Stroyan (1960), Heie (1964b); Greenland: Hille Ris Lambers (1952, 1960); the Faroes: Jørgensen (1932), Heie (1972b). — Britain: Stroyan (1964b, 1966, 1972, 1977); N Germany: Gleiss (1967), F. P. Müller (several papers); Poland: Szelegiewicz (1968b, 1969a). — Europe: Börner (1952), Hille Ris Lambers (1938-1953), Remaudière (1951, 1954), F. P. Müller (1961, 1968, 1969), Léclant (1966, 1967, 1968), Holman (1965, 1971 a.o.), Mier Durante (1978), Ilharco (1960, 1961, 1967-1969, 1973), Szelegiewicz (1968b, 1968c, 1977, 1978a, 1978b), Eastop & Tanasijevic (1968), Semal (1956). — Middle East: Bodenheimer & Swirski (1957), Tuatay & Remaudière (1954). — Far East: Paik (1965), Tao (1961–1970), Higuchi & Miyazaki (1969), Miyazaki (1971). — S Asia: Raychaudhuri (1969 a.o), Robinson (1972), Szelegiewicz (1968a). — USSR: Shaposhnikov (1964). — Mongolia: Holman & Szelegiewicz (1971, 1972), Holman (1972). — Pacific Islands: Essig (1956), Zimmerman (1948). — Australia: Eastop (1966). — New Zealand: Cottier (1953). — Africa: Eastop (1958, 1961), F. P. Müller (1962), v.Harten & Ilharco (1970, 1972, 1976), Ilharco (1969, 1971).—N America: Smith & Parron (1978). — C & S America: Essig (1953a), Smith & Cermeli (1979).

Records from Finland were supplied by O. Heikinheimo, from Norway by Helene Tambs-Lyche and Chr. Stenseth, from Sweden by F. Ossiannilsson and R. Danielsson, and from Denmark by me, including trap data from the suction trap operated at Tåstrup near Copenhagen by the Rothamsted Experimental Station. Records from Germany were supplied by F. P. Müller. V. F. Eastop gave some new information concerning the world distribution.

The colour illustrations were made by Grete Lyneborg on the basis of two drawings of each specimen figured, one based on my own measurements of mounted individuals, the other on my free-hand drawing (with colours) of a live specimen of the same species settled under a stereo microscope. The ink-drawings were made by me after nature, usually on the basis of several measurements of each specimen or appendage, or by means of a drawing device, or they have been taken from papers (the authors of which are mentioned in the legends to the figures).

Acknowledgements

I owe a debt of gratitude for loan or gift of material, and for help to provide distributional information to Mr. Roy Danielsson, Lund, Sweden, Dr. V. F. Eastop, London, Dr. O. Heikinheimo, Vantaa, Finland, Dr. D. Hille Ris Lambers, Bennekom, the Netherlands, Professor Dr. F. P. Müller, Rostock, DDR, Professor Dr. F. Ossiannilsson, Uppsala, Sweden, Mr. J. Reitzel, Lyngby, Denmark, Professor Dr. J. Pettersson, Uppsala, Sweden, Dr. Chr. Stenseth, Vollebekk, Norway, Dr. H. L. G. Stroyan, Harpenden, England, Professor Dr. H. Szelegiewicz, Warszawa, and Dr. Helene Tambs-Lyche, Virum, Denmark (formerly Bergen, Norway).

I thank Mrs. Grete Lyneborg for the colour illustrations and Mr. L. Lyneborg for directions, advice and help in preparing the manuscript.

I express my thanks to the following institutions and foundations for financial support and other facilities: The Carlsberg Foundation, the Danish Science Research Council, the Danish Ministry of Education, Skive Seminarium, the British Museum (Natural History), the Rothamsted Experimental Station, and the Zoological Museum of Copenhagen.

Historical survey

A comprehensive literature on aphids, especially the noxious species, exists, also in the Nordic countries. A survey on the broad outlines of aphid studies in Scandinavia is given here.

In the 18th century Carl von Linné, the Swedish naturalist, issued several editions of his famous "Systema Naturae" with descriptions of the aphid species known at that time based partly on the works of the French entomologist René Antoine de Réaumur. The 10th edition from 1758 is the most important one, with its consistent use of binary animal names being the fix point of zoological nomenclature. More aphid descriptions were added in 1761 and 1767. Most aphids were placed in the genus *Aphis,* a few of them in the genus *Chermes* (viz. *ulmi* and *abietis*), which mostly comprised psyllids. Not only was Linné a systematist and founder of the binary nomenclature, but even a fine observer of biological phenomenons. He discovered that ants fed on the honeydew of aphids and called the aphids the milk cows (vaccae) of the ants.

From the same century the studies of the Swede, Carl DeGeer (1755, 1773), should be pointed out. He described several species and noted that two kinds of females occur within the same species, viz. viviparous females and — late in the year — oviparous females, whose eggs hibernate. A few years earlier Charles Bonnet (1745) in France discovered parthenogenesis of the viviparous females. C. DeGeer described the copulation of males and oviparous females.

The Dane, J. C. Fabricius, a student of Linné, most of his time working in Kiel (then in Denmark), described several aphid species from 1775 to 1803.

From about 1800 to about 1880, only a few Scandinavian entomologists studied aphids, e.g. J. W. Zetterstedt (1828, 1840) in Sweden, who described some new species, and A. G. Dahlbom (1851), who published observations from southern Sweden. Outside the Nordic countries aphid systematics was founded by T. Hartig (1841), who subdivided the genera into two groups, one with a forked media and the other with a simple media, and monographs were published by J. H. Kaltenbach (1843), C. L. Koch (1854–57), and G. B. Buckton (1876–83). G. Passerini (1860, 1863) revised the classification of previous authors. C. H. G. v. Heyden (1837) and F. Walker (1836–72) also should be mentioned.

At the beginning of that period it was discovered that an aphid species could feed on more than one kind of plants. Therefore an aphid found on a new host plant should no longer be described as a new species. Walker discovered the host-alternation in 1847 (migration of *Phorodon humuli* from *Prunus* to *Humulus*) in England.

At the end of the 19th century and beginning of the 20th century several Nordic authors studied the aphids. H. Siebke (1874) published a list of species found in Norway. Others dealt with the biology of the species which were injurious to cultivated plants; in Denmark e.g. J. E. V. Boas studied noxious forest insects, especially aphids on conifers (1890–1924), and S. Rostrup (1897–1912) agricultural pests. In Sweden A. Tullgren (1909–29) published about noxious aphids. He also carried out taxonomic work and gave important contributions to the understanding of the biology of some difficult groups, partly based on experiments with gall-making Pemphigidae. G. Adlerz (1913) investigated root aphids in ants' nests, G. Alm (1915) beech aphids, and O. Gertz (1918) galls. In Norway W. M. Schöyen (1893–1912) dealt with agricultural and horticultural pests. T. H. Hintikka collected aphids for the entomological museum of Helsinki University and studied the Pemphigidae of Finland.

Outside Scandinavia aphids were investigated by entomologists all over the world from the end of the 19th century. The history of aphidology till the middle of the 20th century was published by Essig (1953b) and Cottier (1953). The most famous aphid taxonomists were A. K. Mordvilko (Russia) and C. Börner (Germany), and — since 1931 — first and foremost D. Hille Ris Lambers (the Netherlands).

Among the Scandinavian entomologists who published papers on aphids during the latest decades are: Finland: W. M. Linnaniemi (1915–35), R. Krogerus (1928, 1932), E. Kangas (1931–48), E. Suomalainen (1935), W. Hellén (1935–66), O. Heikinheimo (from 1943), M. Markkula (from 1953), P. Nuorteva (from 1957),

A. Nordman (1959), E. Thuneberg (1960–66), V. Kanervo (1962), J. Rautapää (from 1965); Norway: T. H. Schöyen (1916–49), H. Tambs-Lyche (from 1950), J. Fjelddalen (1964), C. Stenseth (from 1968); Sweden: O. Lundblad (1923–27), O. Ahlberg (1934), E. Wahlgren (1915–62), A. Lindblom (1936–41), F. Ossiannilsson (from 1941), Å. Borg (1949–51), H. Andersson (from 1955), E. Sylvén (1950), B.-O. Landin (1967), J. Lundberg (1963–64), J. Pettersson (from 1968), R. Danielsson (from 1972); Denmark: M. Thomsen (1931–40), C. Stapel (1943–64), P. Bovien (1939 a.o.), K. L. Henriksen (1944), S. G. Larsson (from 1940), Børge Petersen (1959), B. Bejer-Petersen (from 1957), J. Reitzel (from 1970), O.E. Heie (from 1952).

D. Hille Ris Lambers (the Netherlands) and H. L. G. Stroyan (England) described species found in Sweden. A. K. Mordvilko and G. K. Shaposhnikov (the USSR) added to the knowledge of the aphid fauna of East Fennoscandia. In 1971 a suction trap was established at Tåstrup near Copenhagen in connection with a research project carried out by the Rothamsted Insect Survey (at the Rothamsted Experimental Station, Harpenden, England). Some of the results obtained there by Mrs. J. B. Cole and other members of the staff of the Rothamsted Insect Survey are mentioned below.

Systematics

The infraorder Aphidodea is distinguished from other groups within Hemipteroidea by the combination of characters given in the key by Ossiannilsson (1978: 10–11): 1) head without a distinct sclerotized gula, 2) rostrum apparently arising between or behind the fore coxae, 3) tarsi 2-segmented, 4) two pairs of wings present in winged morphs, 5) clavus absent, 6) some morphs without wings. The following distinguishing characters may be added: 7) ocular tubercles with three ommatidia, 8) wings with only one, strongly developed, longitudinal vein (the main vein), 9) a more or less distinctly developed processus terminalis (Fig. 1).

The Aphidodea is subdivided into three superfamilies, viz. the Mesozoic and now extinct Canadaphidoidea, and the still existing Phylloxeroidea and Aphidoidea.

All morphs of Phylloxeroidea are oviparous, while the parthenogenetic females of Aphidoidea are viviparous. This paper does not deal with Phylloxeroidea, only with Aphidoidea. The chief distinguishing morphological and apomorphous character of this superfamily is the siphunculi or siphuncular pores (in some cases reduced or absent).

The number of aphid classifications is about the same as the number of aphid taxonomists. My phylogenetic view on Canadaphidoidea and Phylloxeroidea is published elsewhere (Heie 1980 and 1976a, respectively), and also my view on the phylogenetic tree of Aphidodea as a whole, partly based on fossils (Heie 1967b), but at that time I used Börner's classification from 1952, bound by tradition, although I did not agree with Börner as to 1) the sequence of families (I regard Lachnidae as a relatively yong family), 2) the extent and limitation of Thelaxidae, 3) the inclusion of *Greenidea* in Chaitophoridae (Börner & Heinze 1957), and 4) the classification of Callaphididae. Therefore the classification used in this paper should be explained.

Opinions differ as to the classification of Aphidoidea, while the subdivision of Phylloxeroidea into two recent families, Adelgidae (Chermiden Herrich-Schaeffer in Koch, 1857 = Chermesiens Lichtenstein, 1885 = Adelgidae Börner, 1930) and Phylloxeridae (Phylloxeriden Herrich-Schaeffer in Koch, 1857 = Phylloxeriens Lichtenstein, 1885 = Phylloxeridae Börner, 1930), is old and generally accepted. A third family, the extinct Elektraphididae Steffan, 1968, was added later.

In the present paper the name Aphidoidea has the same meaning as in Shaposhnikov (1964). The name was used in another sense by Baker (1920), Börner (1930, 1952) and several other authors, including Ossiannilsson and me (until 1976), viz. as the name of all aphids (Aphidodea), including Adelgidae and Phylloxeridae. Aphidoidea of Börner corresponds to Aphidina of Burmeister and Aphidodea of Mordvilko, while Aphidoidea in the present sense is the same as Aphidina viviovipara of Börner (who mentions Horvath as the author of that taxon).

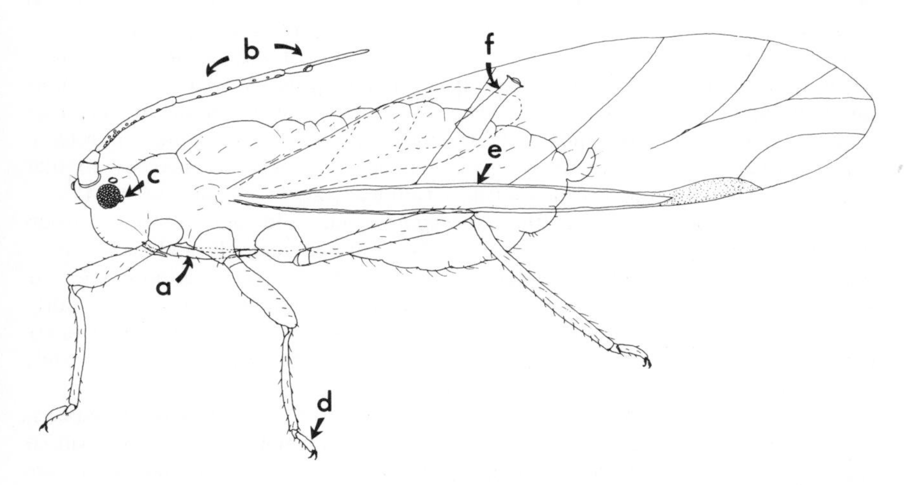

Fig. 1. Distinguishing characters of Aphidoidea. — a = rostrum placed between and behind fore coxae; b = antenna consists of two short basal segments and a thinner flagellum with at most four segments, the ultimate of which consists of a basal part and a thinner apical part, the processus terminalis; c = ocular tubercle (triommatidion) behind the compound eye; d = two tarsal segments; e = wing with only one strongly developed longitudinal vein; f = siphunculus.

The name Aphididae also had several meanings. It may mean all aphids (Aphidodea), the aphids proper (Aphidoidea), Aphidoidea minus the *Pemphigus*-group, or a still smaller part of the Aphidoidea.

Aphidoidea is regarded as one family, Aphididae, by several authors, e.g. Oestlund (1918), Baker (1920), Hille Ris Lambers (1946), and Eastop & v. Emden (1972), while the group is subdivided by other authors into two or more families.

The disagreement as to the classification naturally is not total. As far as I can see, all taxonomists — since the appearance of Kaltenbach's monograph (1843) — distinguished an *Aphis*-group from a *Pemphigus*-group (by different criteria), and present days' taxonomists accept the following unities (usually arranged as subtribes): 1) *Mindarus*-group, 2) *Hormaphis*-group, 3) *Phloeomyzus*-group, 4) *Thelaxes*-group, 5) *Anoecia*-group, 6) *Pemphigus*-group, 7) *Drepanosiphum*-group, 8) *Callaphis*-group, 9) *Chaitophorus*-group, 10) *Greenidea*-group, 11) *Pterocomma*-group, 12) *Aphis*-group, and 13) *Lachnus*-group.

Most of these groups were placed at the same level (tribes though with names ending in -ina instead of -ini) by van der Goot (1913), who regarded Aphidoidea as a subfamily (Aphidinae) of Aphididae (= Aphidodea in the present sense).

Authors do not agree as to 1) the number of such groups (some of them may be subdivided), 2) the sequence (nobody used the sequence mentioned above), 3) the rank(s), and 4) the lumping of these groups into larger groups of Aphidoidea.

The disagreement is due to the difficult recognition of the evolutionary trends, which may not be equal in all groups. It is difficult to decide what are plesiomorphous or apomorphous characters, partly because parallel evolution has involved a mosaic distribution of apomorphous characters (e.g. some wing characters), partly because the trend in many groups is one of progressive neoteny so that geologically young groups lost some advanced adult characters (cf. Richards 1964).

While some old groups as the *Hormaphis*-group and the *Pemghigus*-group specialized long ago and became strongly fixed, morphologically and ecologically, but retained some primitive (plesiomorphous) characters, other groups, e.g. the *Aphis*-group and the *Lachnus*-group, probably remained little-specialized until fairly recently they occupied a number of new ecological niches. The *Aphis*-group today is by far the group richest in species, much larger than the *Pemphigus*-group, while the situation was opposite in the early Tertiary.

Young groups may appear more primitive than old ones in several respects, because their success depended on their ancestors' remaining at a little specialized stage, which meant a great evolutionary plasticity. Apomorphous characters are not lacking in these groups; the autapomorphous characters of e.g. the *Aphis*-group are few, but important all-over-improvements for survival in a changing world, whereas they are numerous, but mostly evolved through reduction and host-specific adaptation in the "old" groups. Eventually the flora changed, and the specialized groups declined or became extinct, while the slightly specialized groups increased. "The phylogeny suggested by fossil Hemiptera should be studied against the background provided by palaeobotany", as said by Eastop (1966: 401), and this is correct, but the new host plants acquired by species of expanding aphid groups may as well be "old" plants as "young" plants. Association with "old" plant groups does not always mean that these aphid genera are "old". Eastop regards the *Lachnus*-group as an old group because most recent species are associated with coniferous trees, and also because most morphological characters apparently are plesiomorphous. This view is shared by most other authors. Arguments for the opposite opinion were given by Mackauer (1965) and Heie (1967b: 229, 252). The *Lachnus*-group is in agreement with this view treated in the last volume.

Disagreement as to the rank of the subgroups depends on individual judgments. I have treated the main groups as families following Börner (1952) and Shaposhnikov (1964), though the families are less distinctly defined than those of Phylloxeroidea, probably due to a later branching of the phylogenetic tree in Aphidoidea than in Phylloxeroidea. For several years this has been the usual procedure in Scandinavia and some neighbouring countries, and has also been accepted by several entomologists in other parts of the world.

Another reason to give family rank to the higher groups is the need for three categories of lower rank, viz. subfamily, tribe and subtribe (e.g. in Aphididae: Aphidinae — Aphidini — Rhopalosiphina and Aphidina).

With reference to recent phylogenetic theories advanced by Hille Ris Lambers (1964, 1968), Ilharco (1964), and Heie (1967b) I do not accept Börner's Classification of Thelaxidae, Chaitophoridae, and Callaphididae, and not either his subdivision of Aphididae into five subfamilies.

Thelaxidae sensu Börner is a paraphyletic group and should be subdivided into several "old" families. Shaposhnikov (1964) removed the *Mindarus*-group (Mindaridae), the *Anoecia*-group (Anoeciidae), and the *Phloeomyzus*-group (Phloeomyzidae) from Thelaxidae, but not the *Hormaphis*-group, which is the most incongruous group included in Thelaxidae by Börner. It should also be removed. Hormaphididae shows several autapomorphous characters, and shares some apomorphous characters with other groups, not only with the *Thelaxes*-group, but also with the *Pemphigus*-group. Attempts to place the *Hormaphis*-group together with any other group were never generally accepted. Eastop in 1961 regarded it as a tribe within Thelaxinae, in 1966 as a subfamily within Pemphigidae, and in 1972 as a subfamily within Aphididae (Aphididae in 1972 was used in the sense of Aphidoidea by Shaposhnikov and me; the *Pemphigus*-group was also regarded as a subfamily, while it was regarded as a family in 1966).

Szelegiewicz (1978b) regarded the *Hormaphis*-group and the *Anoecia*-group as families, but the *Thelaxes*-, *Phloeomyzus*-, and *Mindarus*-groups as tribes of one subfamily, Thelaxinae, within Phyllaphididae (= Callaphididae of Börner).

Chaitophoridae is a well-defined monophyletic group if the *Greenidea*-group is left out. Callaphididae sensu Börner is not a monophyletic, but a paraphyletic group, and Börner's subdivision of it — mainly based on the morphology of the first instar nymphs — was criticized by Ilharco (1964) and others.

The differences between Chaitophoridae and Callaphididae are small, and they should either be lumped into one family or regarded as more than two families, to make classification agree with phylogenetic considerations. Mackauer (1965) and Hille Ris Lambers (1968) believe — on the basis of the systematics of aphid parasites and of the *Neuquenaphis*-group, respectively — that the *Drepanosiphum*-group, together with Chaitophoridae, constitute the sister-group of Callaphididae sensu Börner minus the *Drepanosiphum*-group, which view also Heie accepted (1967b: 250–251) — on the basis of studies of fossil aphids.

If this is correct, either Chaitophoridae should be reduced to a subfamily of Callaphididae as in the classification of Bodenheimer & Swirski (1957), or the *Drepanosiphum*-group be removed from Callaphididae as in the classification by v.d. Goot (1913), and given family rank. I prefer the former solution. In either case, the consequence will be a subdivision of Chaitophoridae + Callaphididae sensu Börner (Drepanosiphidae is used as the name of this group by the present author) into a minimum of three groups:

I. Drepanosiphinae Herrich - Schaeffer in Koch, 1857
 Neophyllaphidini Takahashi, 1921
 Paoliellini Takahashi, 1930
 Spicaphidini Essig, 1963 (Neuquenaphidini Heie, 1967)
 Palaeosiphonini Heie, 1967 (fossil)
 Israelaphidini Ilharco, 1961
 Drepanosiphini Herrich-Schaeffer in Koch, 1857

II. Phyllaphidinae Herrich-Schaeffer in Koch, 1857
 Phyllaphidini Herrich-Schaeffer in Koch, 1857
 Macropodaphidini Aizenberg, 1960
 Saltusaphidini Baker, 1920

III. Chaitophorinae Mordvilko, 1908
 Chaitophorini Mordvilko, 1908
 Siphini Mordvilko, 1928

Some comments on the nomenclature shall be added. The oldest available family-group names are Drepanosiphidae (originally spelled Drepanosiphiden), type genus *Drepanosiphum* Koch, 1855, and Phyllaphididae (originally spelled Phyllaphiden), type genus *Phyllaphis* Koch, 1857, published by D. Herrich-Schaeffer in C. L. Koch (1857: vii).

A third name created by Herrich-Schaeffer at the same time, Callipteridae (spelled Callipteriden), is unavailable because its nominal type genus, *Callipterus* Koch, 1855, is a junior homonym of *Callipterus* Agassiz, 1846 (not an aphid). The replacement name of *Callipterus* Koch is *Callaphis* Walker, 1870.

Mordvilko (1908) placed *Drepanosiphum* in Aphidina, *Callipterus* and *Phyllaphis* in Callipterina, and — as the first one — formed a family-group name with *Chaitophorus* Koch as the nominal type genus, viz. Chaitophori (belonging to tribe Aphidina).

The tribes Drepanosiphina, Callipterina, and Chaitophorina of van der Goot (1913) correspond to the subfamilies mentioned above. He did not include *Pterocomma* Buckton (*Cladobius* Koch) which had been placed in Phyllaphiden by Herrich-Schaeffer, in Chaitophori by Mordvilko, and he did not include the *Lachnus*-group, which had been placed in Callipterina by Mordvilko.

Baker (1920) apparently as the first one placed *Drepanosiphum, Callipterus, Phyllaphis,* and *Chaitophorus* in one tribe, Callipterini, but he also included *Pterocomma* Buckton, and each of them were treated as nominal type genera of subtribes. Bodenheimer & Swirski (1957) followed Baker, but omitted *Pterocomma* from the group, which they gave family rank (Callipteridae).

Börner (1930) placed *Drepanosiphum, Callipterus,* and *Phyllaphis* in Callipterini and *Chaitophorus* in Chaitophorini, but he later (1952) changed the names of these groups into Callaphididae and Chaitophoridae. The nominal type genus of the former family-group name shall, however, be either *Drepanosiphum* (as in the classification of Eastop in Eastop & v. Emden 1972) or *Phyllaphis* (as in the classifications of Szelegiewicz 1978 and several others) because the law of priority (Article 23 of the International Code of Zoological Nomenclature) applies to family-group names, and Callaphididae is a younger name than Drepanosiphidae and Phyllaphididae. The name of the second subfamily above shall therefore be Phyllaphidinae instead of Callaphidinae.

The family consisting of Drepanosiphinae, Phyllaphidinae, and Chaitophorinae can be called either Drepanosiphidae Herrich-Schaeffer or Phyllaphididae Herrich-Schaeffer depending on the determination of the first reviser. Drepanosiphidae is hereby adopted as the valid name of the family. It has been taken into consideration that the name Phyllaphididae has been used more frequently as the name of part of the family (Drepanosiphinae + Phyllaphidinae) than the name Drepanosiphidae.

Aphididae in this paper is used in the sense of Börner (1952, not 1930), but his subdivision of this family into five subfamilies was long ago invalidated by demonstration of the so-called Rosaceous series by Hille Ris Lambers (1939a: 76–77, 1950: 143). It should be subdivided into two subfamilies only, viz. Pterocommatinae Mordvilko and Aphidinae Mordvilko. Anuraphidinae, Myzinae, and Dactynotinae of Börner are grades instead of clades and should be placed in one tribe, Macrosiphini, within Aphidinae (Eastop 1961). Aphidinae of Börner is regarded as another tribe, Aphidini.

The classification proposed in accordance with these remarks is given below together with the corresponding group-names of the classifications used by 1) Börner (1952) and Börner & Heinze (1957), 2) Shaposhnikov (1964), and 3) Eastop (1972). The numbers give the sequence of the main groups of each classification.

Baltichaitophorinae n. subfam. is added: Type genus: *Baltichaitophorus* Heie, 1967 (fossil, Baltic amber); diagnosis: Processus terminalis, siphunculi and cauda well-defined as in the other subfamilies of Aphididae, but compound eyes of apterae 3-lensed. Distinct from Cervaphidinae by lacking abdominal processes, and from

Traminae by having hind tarsi about as long as fore and middle tarsi. The recent genus *Parachaitophorus* Takahashi, 1937, is included in the new subfamily (see also Heie 1967b: 181, 247).

Classifications

Proposed classification	*Börner*	*Shaposhnikov*	*Eastop*
APHIDOIDEA	APHIDINA VIVIOVIPARA	APHIDOIDEA	APHIDIDAE
	5. Thelaxidae		
1. Mindaridae	Thelaxinae: Mindarini partim	3. Mindaridae	(not mentioned)
2. Hormaphididae	Hormaphidinae	6b. Thelaxidae: Hormaphidinae	8. Hormaphidinae
3. Phloeomyzidae	Thelaxinae: Mindarini partim	5. Phloeomyzidae	(not mentioned)
4. Thelaxidae	Thelaxinae: Thelaxini	6a. Thelaxidae: Thelaxinae	(not mentioned)
5. Anoeciidae	Anoeciinae	4. Anoeciidae	7. Anoeciinae
6. Pemphigidae Eriosomatinae Pemphiginae Fordinae	6. Pemphigidae Eriosomatinae Pemphiginae Fordinae	1. Pemphigidae Eriosomatinae Pemphiginae Fordinae	9. Pemphiginae Eriosomatini Pemphigini Fordini
7. Drepanosiphidae			
Drepanosiphinae	3. Callaphididae: Phyllaphidinae partim	7. Callaphididae: Callaphidini partim	3. Drepanosiphinae: Drepanosiphini partim
Phyllaphidinae	Phyllaphidinae partim + Callaphidinae + Therioaphidinae + Saltusaphidinae	Callaphidini partim + Saltusaphidini	Drepanosiphini partim + Saltusaphidini
Chaitophorinae	2. Chaitophoridae Chaitophorinae + Siphinae	8. Chaitophoridae	2. Chaitophorini
8. Greenideidae Greenideinae Cervaphidinae	Trichosiphoninae	(not mentioned)	6. Greenideinae Greenideini Cervaphidini

9. Aphididae	4. Aphididae	9. Aphididae	
Baltichaitophorinae (new)			
Pterocommatinae	Pterocommatinae	Pterocommatinae	4. Pterocommatinae
Aphidinae		Aphidinae	5. Aphidinae
Aphidini	Aphidinae	Aphidini	Aphidini
Macrosiphini	Anuraphidinae + Myzinae + Dactynotinae	Macrosiphini	Macrosiphini
10. Lachnidae	1. Lachnidae	2. Lachnidae	1. Lachninae
Lachninae	Lachninae	Lachnini	Lachnini
Cinarinae	Cinarinae	Cinarini	Cinarini
Traminae	Traminae	Tramini	Tramini

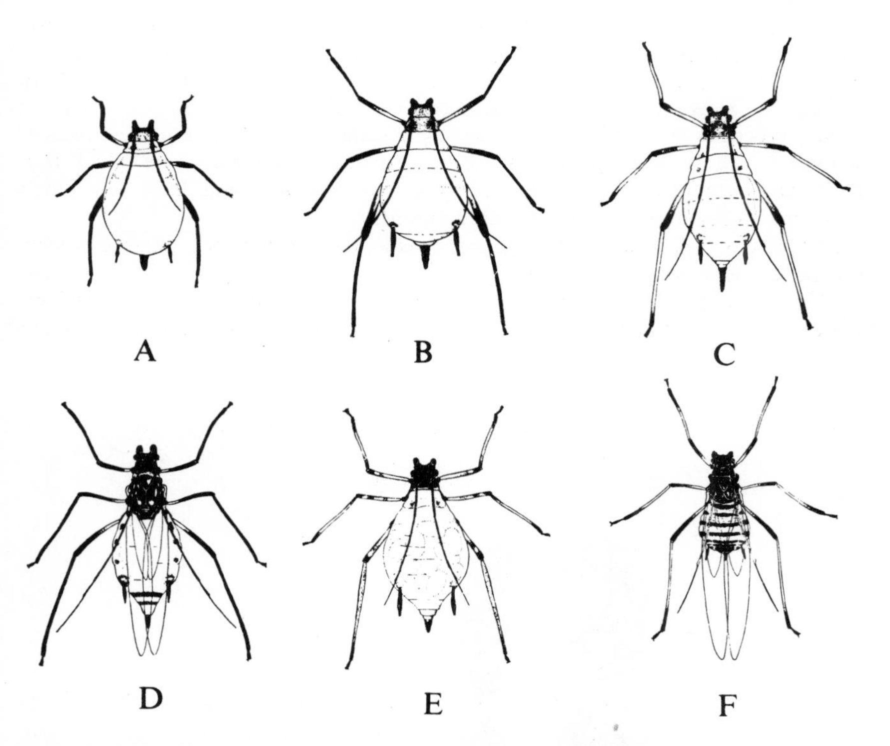

Fig. 2. Various morphs of *Megoura viciae* Buckt. – A: fundatrix; B and C: apterous viviparous females from spring and summer, respectively; D: alate viviparous female; E: oviparous female; F: male. (After Lees).

Polymorphism

Aphids are usually considered a ''difficult'' group, taxonomically and ecologically. One reason is their polymorphism, the occurrence of several kinds or morphs of adult individuals within the same species. Some species may show generations with highly varying individuals. On the other hand species exist which are so similar that they cannot be separated by representatives of a single morph.

There can be more than ten morphs of one species (e.g. in some *Periphyllus* spp.) or less than five, but usually the following five morphs occur (cf. Fig. 2):

1) Fundatrix (stem mother), parthenogenetic female from a fertilized egg; in Aphidoidea always viviparous.

Fundatrix hatches from an overwintering egg in spring and becomes the stem mother of generations of parthenogenetic females of the same genotype (Fig. 35). It is generally apterous and fairly similar to the apterous viviparous females of subsequent generations though usually with a thicker body, shorter antennae, shorter processus terminalis, shorter legs, and sometimes with more or less reduced siphunculi. In a few species of Drepanosiphidae fundatrices are alate (e.g. *Drepanosiphum platanoidis* and *Euceraphis punctipennis*), and in a few others some fundatrices are alate, and some are apterous.

2) Apterous viviparous parthenogenetic female, the daughter of another parthenogenetic female (fundatrices are then excluded).

This is the best described morph because it is found more frequently than any other morph of most species, often in colonies. In many little known species this

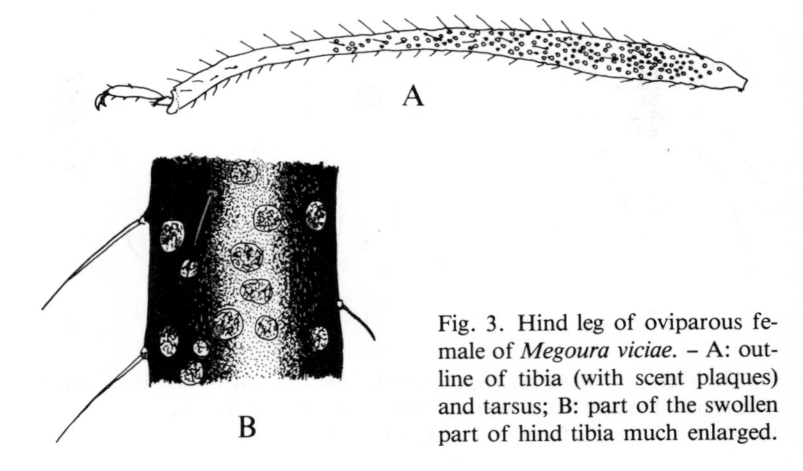

Fig. 3. Hind leg of oviparous female of *Megoura viciae*. – A: outline of tibia (with scent plaques) and tarsus; B: part of the swollen part of hind tibia much enlarged.

morph is the only one ever described. In most species it occurs throughout the summer. It is a morph specialized for large reproduction, while the alate parthenogenetic female (3) is specialized for dispersal. Apart from the ability of reproduction the adult apterous female is very similar to the immature individual.

3) Alate viviparous parthenogenetic female, the daughter of another parthenogenetic female.

It may occur in the same generations as the apterous viviparous female (2). Alate viviparous females (alatae) and apterous viviparous females (apterae) consequently may be sisters. In some species the alate viviparous morph only occurs in some generations, either as the only morph, or together with the apterous viviparous morph. In other species alatae appear in any generation influenced by certain environmental factors. The alate viviparous female is the only parthenogenetic female in a few species, while it is rare or unknown in others. The alata usually bears fewer young than the apterous female.

Intermediate forms between apterae and alatae sometimes occur. Such specimens in some cases are abnormal, so-called intermediates, with incomplete wings, in other cases so-called alatiform apterae. Alatiform apterae are common (normal) in some species, e.g. some *Dysaphis* spp. They are apterous females with some features characteristic of the alate female, e.g. the ocelli, a larger number of secondary rhinaria, and strong pigmentation and sclerotization of thorax, or at least with one of these attributes.

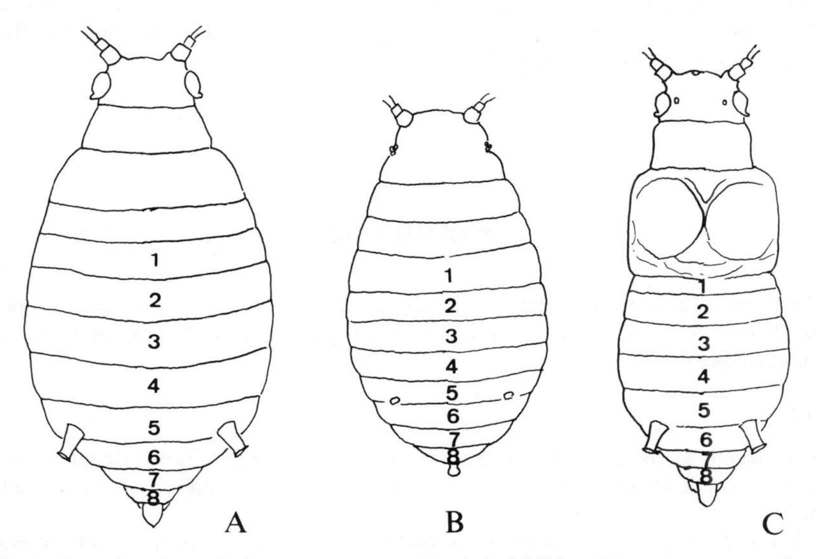

Fig. 4. Segmentation of A: apterous morph of Aphididae; B: apterous morph of Thelaxidae; C: alate morph of Aphididae; 1–8 = abd. segm. I–VIII.

4) Oviparous female (ovipara), able to reproduce after mating; the daughter of a parthenogenetic female; the mother of fundatrix.

This morph is apterous in all Scandinavian species, and in most aphid species. It is the only oviparous morph of Aphidoidea. It usually resembles the apterous viviparous female, but in some species it is dwarfish and more or less reduced. In most aphids it is distinguished by scent plaques on the more or less swollen hind tibiae (Fig. 3); the males are attracted by the scent (Pettersson 1970).

5) Male, the son of a parthenogenetic female; the father of fundatrix; copulates with the oviparous female.

The male is alate or apterous. In some species both alate males and apterous males occur. The male usually resembles the alate viviparous female, also when it is apterous, e.g. as regards pigmentation and sensory organs. It is distinguished from the females by the genitalia (Fig. 33B) and the greater number of rhinaria. In some species the males are dwarfish and more or less reduced.

Oviparae (4) and males (5) are called sexuales or the sexualis-generation. Their mothers belong to morphs 2 and/or 3. The mothers are called sexuparae, if they bear both oviparae and males, gynoparae if they bear oviparae only, and androparae if they bear males only. In many species the same individual may bear parthenogenetic females as well as sexuales.

Host-alternating (dioecious, migrating) species with apterous males (Fig. 43) have alate sexuparae (or androparae), while those with alate males usually have apterous androparae. Sexuparae and gynoparae of host-alternating species are always alate because the oviparous females are apterous and cannot migrate from the secondary host to the primary host themselves.

Morphology

The terms used in the descriptions and keys are in accordance with modern European aphidology. For the guidance of non-aphidologists the most important terms are explained below.

Body

Body length is measured from the anterior edge of head (frons) (from anterior edge of frontal tubercles if present) to apex of cauda. Most aphids are 1.5–3.5 mm long.

The segment limits of the body are more or less conspicuous. The limit between head and prothorax is invisible in immature individuals and apterae of some groups (Mindaridae, Thelaxidae a.o.) (Fig. 4 B).

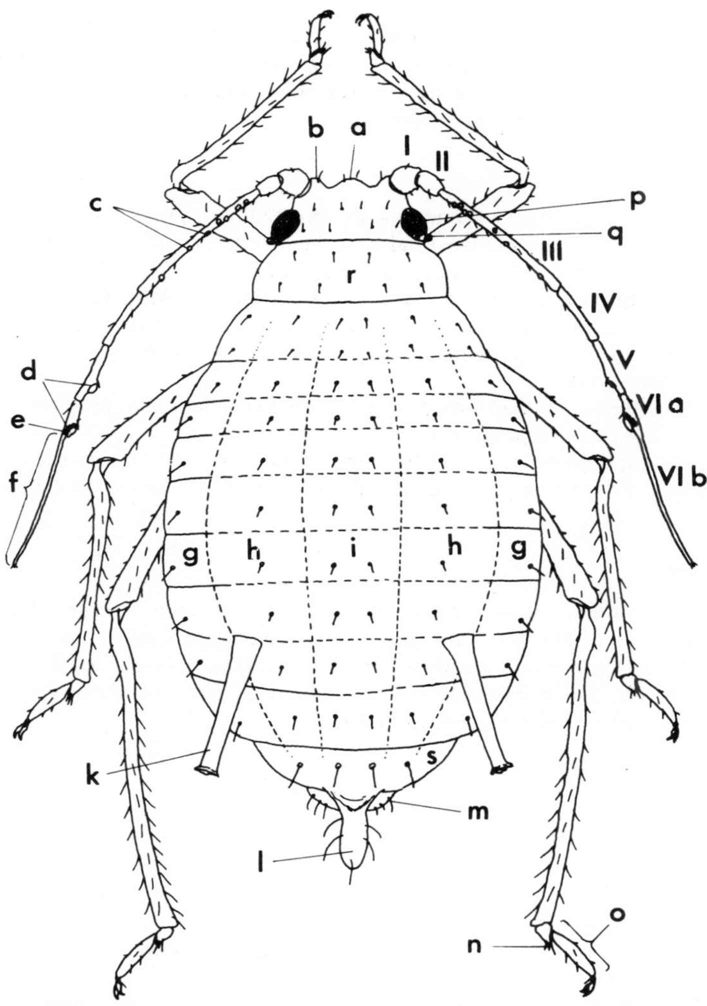

Fig. 5. Apterous aphid, viviparous female. – a = median frontal tubercle; b = lateral frontal tubercle; c = secondary rhinaria; d = primary rhinaria; e = accessory rhinaria (close to the primary rhinarium of ant. segm. VI); f = processus terminalis; g = marginal areas of dorsum; h = pleural areas of dorsum; i = spinal area of dorsum; k = siphunculus; l = cauda; m = anal plate; n = hairs on first tarsal segment; o = second tarsal segment; p = compound eye; q = ocular tubercle (triommatidion); r = pronotum; s = abd. segm. VIII; I–V = ant. segm. I–V; VIa = basal part of ultimate ant. segm.; VIb = processus terminalis.

The thoracic and abdominal segments are usually fairly uniform as to sclerotization and pigmentation in apterous females, while alate aphids (Fig. 4 C) — and also apterous males — have stronger sclerotization and usually darker pigmentation of meso- and metathorax than of abdomen.

Hairs

The basic dorsal pattern is formed by six longitudinal rows of hairs or groups of hairs, viz. spinal hairs (situated fairly close to the dorsal mid-line, Fig. 5 i), pleural hairs (between the spinal and marginal hairs, Fig. 5 h), and marginal hairs (Fig. 5 g), arranged in a transverse row on each segment. Head and thoracic segments I–II have more hairs, and the posterior abdominal segment (VIII) has fewer. Some of the longitudinal rows may be absent. In many species superpernumerary hairs occur.

The hairs (setae, bristles) may be simple, spatulate, flabellate, mushroom-shaped, ramifying (forked), or capitate (knobbed) (Fig. 6).

In the descriptions the length of the hairs on antennae, frons, and body segments is compared with the length of the basal diameter of ant. segm. III (abbreviation: III bd.).

Wax glands

Wax glands are common in aphids (Fig. 7). In some species, e.g. within Pemphigidae, the wax pores are placed close together on well-defined wax gland plates. At the primitive state these plates are arranged in six longitudinal rows (spinal, pleural, and marginal rows). At the derived state the plates may be fewer due to reduction, or some may be subdivided.

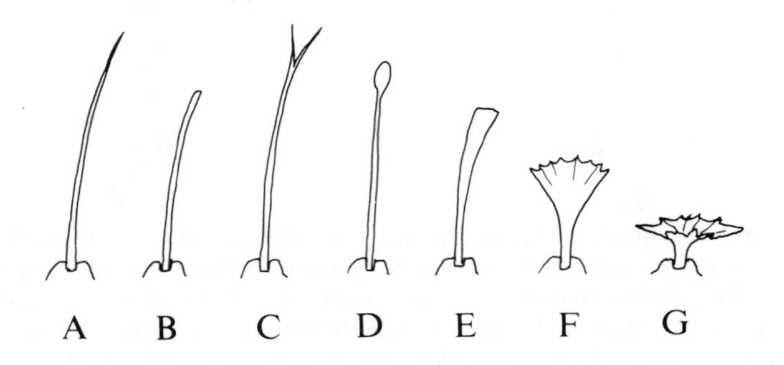

Fig. 6. Various shapes of hairs. – A: simple, pointed; B: simple, blunt; C: ramifying; D: capitate; E: spatulate; F: flabellate or fan-shaped; G: mushroom-shaped.

Wax gland plates occur in Mindaridae, Pemphigidae, and some Drepanosiphidae (*Phyllaphis* a.o.) (Figs. 85, 127 A-E, 207). They are often better developed in apterae than in alatae. The wax is produced as woolly threads, sometimes united into fluffs or bands, or as powder consisting of short pieces of threads.

Unicellular wax glands occur in some aphid species, for instance in Aphididae and Lachnidae. They are differentiated hypodermis-cells scattered all over the body or on certain body parts; the pores are almost invisible. These glands produce a dustlike coating of wax powder making the aphids look greyish and dull.

Tubercles

Wartlike, globular, conical, or finger-shaped tubercles may occur in various positions on the head, thorax, and abdomen. They are called processes if they are fairly large and more or less finger-shaped.

Stigmal pori

Aphids generally have nine pairs of spiracles or stigmal pori, viz. two pairs on thorax and one on each of the seven anterior abdominal segments (Fig. 8). They are usually placed beside small sclerites (stigmal plates) and may be circular, subcircular, or reniform. They are sometimes more or less covered by a small anterior shelter, the operculum.

Head

The head is bent downward along a sharp edge called frons (not homologous with the anatomical frons) so that the base of rostrum is situated on the posterior part of the underside of the head (Fig. 9). The fore-head, which is turned downward, consists of post- and anteclypeus and labrum.

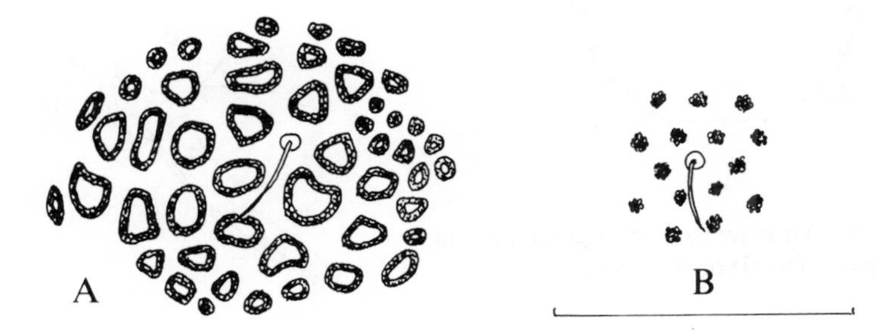

Fig. 7. Wax glands of apterae of two species of Drepanosiphidae. – A: dorsal wax gland plate of *Phyllaphis fagi* (L.); B: part of dorsal cuticle with wax gland pores of *Thripsaphis (Trichocallis) cyperi* (Wlk.). (Scale 0.1 mm).

More or less well-developed frontal tubercles may occur on the frontal edge, viz. one pair of lateral frontal tubercles at the antennal bases, and sometimes also a median frontal tubercle in the middle of the frontal edge between the lateral tubercles. The angle between the inner margins of the lateral tubercles is of taxonomic importance in genera within the Macrosiphini, in which these tubercles usually are particularly well developed. The frontal tubercles are said to be parallel, converging, or diverging, when the inner margins are parallel, converging, or diverging (in forward direction), respectively (Figs 10–12). Sometimes the lateral tubercles carry finger-shaped processes, e.g. in *Phorodon* (Fig. 14).

Eyes

Alate aphids have two large compound lateral eyes and three ocelli. An ocular tubercle with three ommatidia (the triommatidion) occurs at the posterior margin of either eye (Fig. 15 A). It is rarely displaced to the ventral margin of the compound eye, or absent.

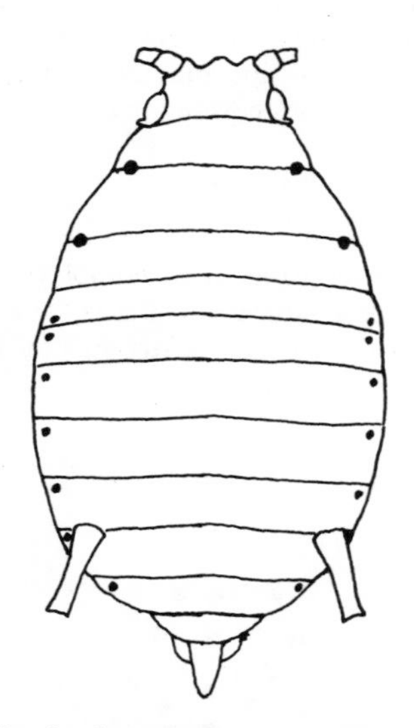

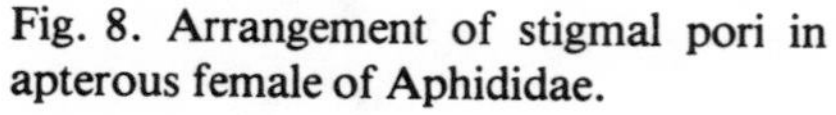

Fig. 8. Arrangement of stigmal pori in apterous female of Aphididae.

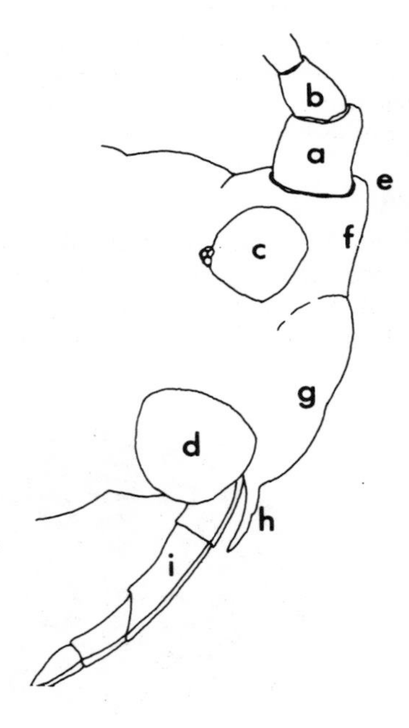

Fig. 9. Outline of head seen from the right side. – a, b = ant. segm. I and II; c = eye; d = position of fore coxa; e = frontal edge (usually called frons); f = anatomical frons; g = lower part of the fore-head (clypeus); h = labrum; i = labium (rostrum).

The triommatidia are the only eyes in newborn nymphs of some groups (Mindaridae, Hormaphididae, Thelaxidae, Anoeciidae, Pemphigidae, and Traminae). In most of these aphids the triommatidia remain the only eyes of all instars of apterae (Fig. 15 B), including the adult stage, unless the apterae are alatiform (p. 23), while a multifacetted eye develops between the triommatidion and the antennal base on either side of the head in nymphs of the alate morph.

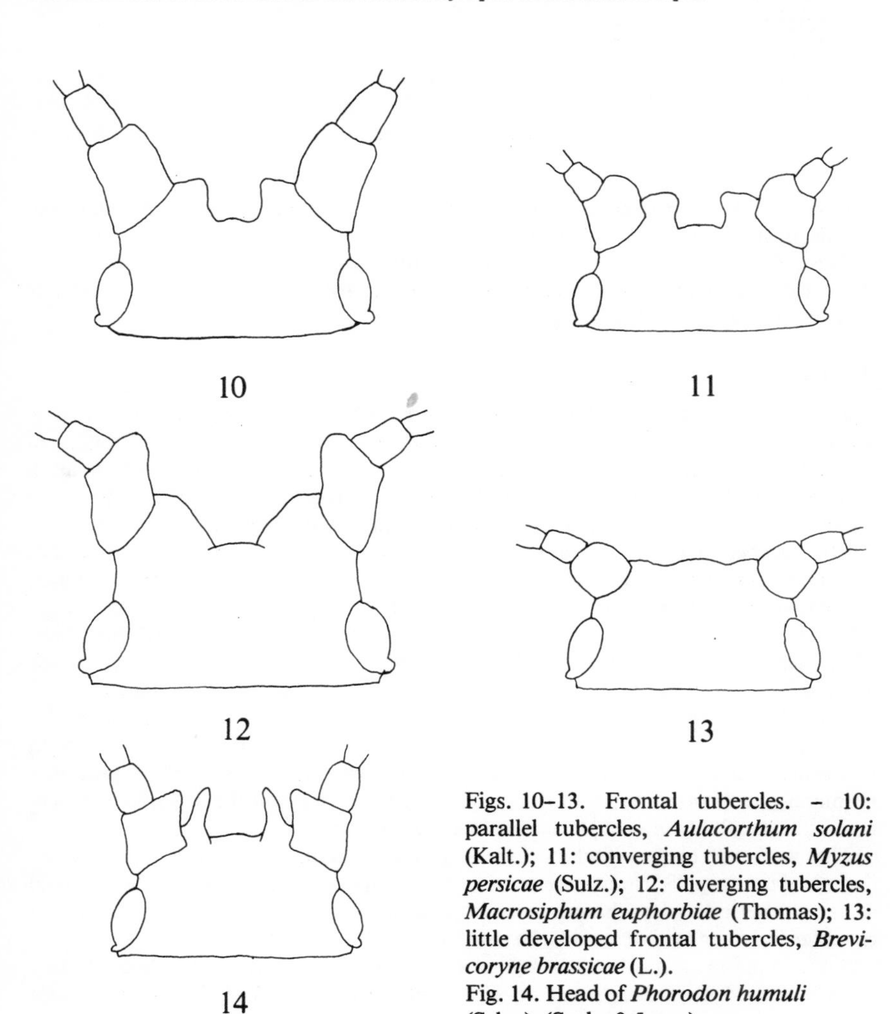

Figs. 10–13. Frontal tubercles. - 10: parallel tubercles, *Aulacorthum solani* (Kalt.); 11: converging tubercles, *Myzus persicae* (Sulz.); 12: diverging tubercles, *Macrosiphum euphorbiae* (Thomas); 13: little developed frontal tubercles, *Brevicoryne brassicae* (L.).
Fig. 14. Head of *Phorodon humuli* (Schr.). (Scale: 0.5 mm).

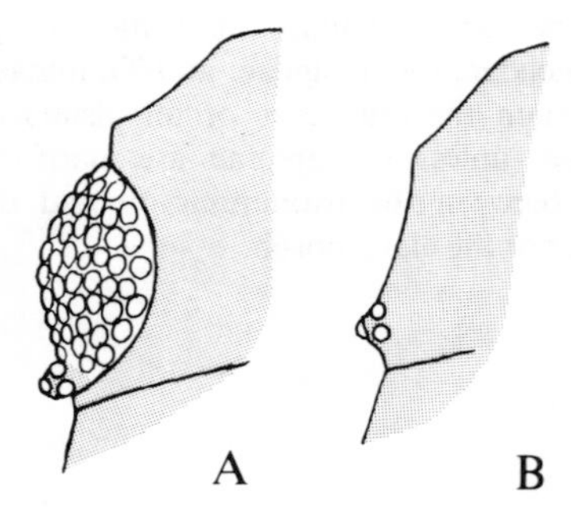

Fig. 15. Two kinds of lateral eyes. – A: compound eye with ocular tubercle (triommatidion); B: triommatidion only.

Large compound eyes are developed before birth in all other groups of Scandinavian aphids, in apterous as well as in alate morphs.

The lateral ocelli are placed close to the inner margins of the compound eyes, the median one on frons. Ocelli are absent in nymphs and in adult apterous females, unless they are alatiform (p. 23).

Antennae

The maximum, and also the normal, number of antennal segments is six. The antennae may be shorter or longer than the body. They are relatively longer in alate than in apterous females of the same species, and longer in males than in females. The basal segments (I and II) are short and relatively thick. The distal segments (III–VI) are longer and thinner and constitute the flagellum.

The number of antennal segments is less than six in young nymphs, sometimes also in adults. This is due to the absence of one or more joints (segment limits) from that part of flagellum which corresponds to segments III-V of the normal antenna. Growth proceeds in nymphs mainly in the proximal part of flagellum, while the two basal segments and the ultimate segment assume the final shape and almost adult proportions relatively early during the ontogenetic development.

The ultimate segment consists of a basal part (called VIa in 6-segmented antennae) about as thick as the penultimate segment, and a thinner apical part, called processus terminalis (VIb in 6-segmented antennae) (Fig. 5 f). The length of processus terminalis, in proportion to the length of the basal part of the ultimate segment and to the length of segment III, is important to taxonomy.

The antennae carry olfactory organs called rhinaria. Seen in the microscope they stand out as pale, circular or transverse oval spots or half rings on the antennal surface. The individual rhinarium consists of a thin membrane surrounded by a cuticular frame or rim, which may be hairy (Fig. 16 E).

Some rhinaria occur both in nymphs and in adults. They are called primary rhinaria (Fig. 5 d). Others occur in adults only. They are called secondary rhinaria (Fig. 5 c).

Two primary rhinaria are present on either antenna, one close to the apex of the penultimate segment and the other at the apex of the basal thick part of the ultimate segment. The distal border of the latter rhinarium by definition is the basal limit of processus terminalis. A cluster of fairly small so-called accessory rhinaria occurs close to the primary rhinarium of the ultimate segment (Figs. 5 e, 17).

Also the secondary rhinaria are placed on flagellum, primarily on that part which had the more conspicuous growth after birth, most frequently on the basal half of segment III.

Alate females have more secondary rhinaria than apterous females of the same species, and males normally have more rhinaria than females. The number is not constant within one morph, but varies within a certain range characteristic of the species. Secondary rhinaria are absent in the apterous viviparous females of several species. Males of host-alternating species within Aphididae carry more secondary rhinaria than males of related species without host-alternation.

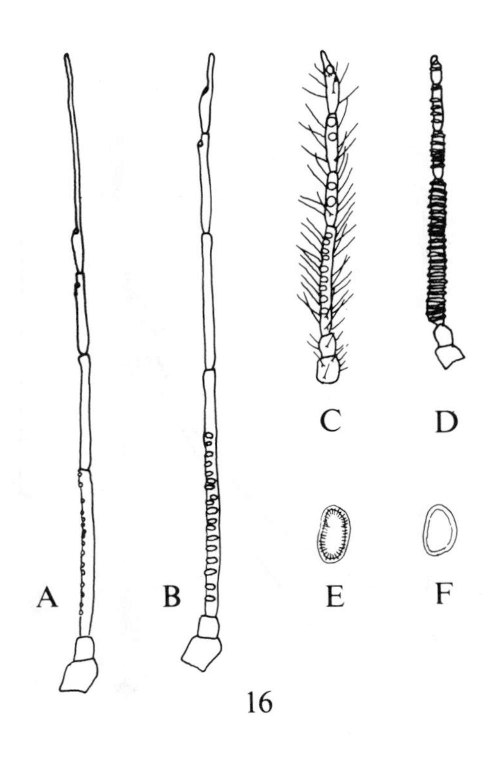

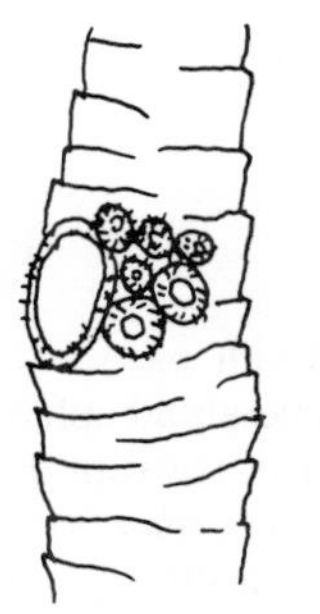

Fig. 17. Primary rhinarium and accessory rhinaria on ant. segm. VI of *Monaphis antennata* (Kalt.).

Fig. 16. A–D: antennae of alate viviparous females. – A: *Myzus,* with circular rhinaria; B: *Betulaphis* and C: *Cinara,* both with transverse oval or subcircular rhinaria; D: *Eriosoma*, with ringlike rhinaria. – E, F: primary rhinaria, surrounded by a ring of short hairs (E) or without such a ring (F).

Mouth parts

The mouth parts are sucking and consist of the rostrum, which is the 4-segmented labium, and four stylets, which are the mandibles and the maxillae.

Segment I of rostrum is usually relatively soft (slightly sclerotized). Its distal end surrounds the basal end of segment II, which is long and rigid. Labium shortens during feeding as a telescope because segment II can be pushed back into segment I. Segment III is short. Segment IV (the apical segment of rostrum) is more or less triangular, acute or obtuse, and carries three pairs of apical hairs (Fig. 18 A, i) and usually one pair of short basal hairs (Fig. 18 A, g). Accessory hairs occur between them in adults (Fig. 18 A, h). The number of these hairs are taxonomically important.

The apical segment of rostrum may be subdivided into two parts, the ultimate being shorter, narrower, and often much darker than the penultimate (Fig. 18 C).

Labium is gutter-shaped, with a deep groove along the front side. This groove contains the stylets. It is almost closed except near the base, where it is covered by the labrum, which is narrow, with concave inner face.

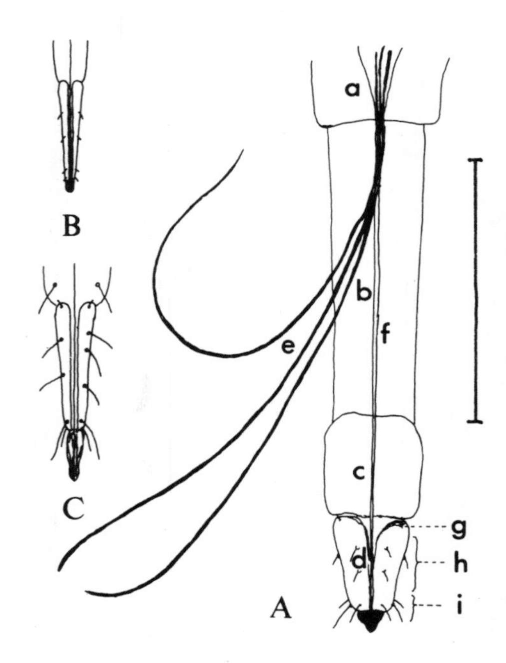

Fig. 18. Mouth parts. – A: rostrum of a generalized aphid; B: apical segment of rostrum of *Macrosiphoniella hillerislambersi* Ossiannilsson (not distinctly subdivided into two parts); C: apical segment of rostrum of *Cinara pilicornis* (Hartig) (subdivided into two parts, often called segm. IV and V). – a, b, c = segments I–III of rostrum; d = apical segment of rostrum; e = stylets (the median stylets are locked together so that only three "threads" are visible, here pushed out from the labial groove); f = labial groove; g = basal hairs of apical segment; h = accessory hairs; i = apical hairs (hairs on segm. I–III omitted). (Scale 0.5 mm).

The lateral (and anterior) stylets are the mandibles. The median (and posterior) stylets are the maxillae, or rather the laciniae of the maxillae. The inner faces of the median stylets are locked together so as to form two channels or tubes, an anterior food channel for uptake of plant sap (Fig. 19 c), and a posterior narrower channel for transport of saliva into the plant tissue (Fig. 19 d). In mounted aphids (see p. 69) the stylets look like three thin threads (one thread consists of two stylets) which during preparation are often pushed out of the labial groove (Fig. 18 A, e). A live aphid keeps the stylets inside the rostrum and the fore-head. when it is not feeding. Then the rostrum lies along the venter, the tip pointing backward (Fig. 1 a).

When an aphid is going to feed, it first lowers the rostrum into a vertical position at right angles to the surface of the plant. The apex of rostrum can be pushed hard against the plant surface because rostrum arises from the posterior part of the ventral side of the head supported by the fore coxae. Muscles fixed to the basal parts of the stylets inside the head pull the stylets down into the plant tissue by small jerks. One mandible opens a passage for the fused median stylets which follow the mandible, either simultaneously (Fig. 20 C) or after a short while (Fig. 20 D-E). In the former case the stylets are plunged straight into the tissue. In the latter case they will curve. Then the other mandible proceeds. After shortening of the rostrum by telescopelike pushing of one segment into another the bases of the stylets are forced into the fore-head again. Then the muscles mentioned above can pull the stylets down again, and so on.

The stylets do not slide back and out of the plant tissue in the pauses because the apical segment of rostrum keeps them in the labial groove just like a pair of tweezers. Tiny transverse ribs at the tips of the mandibles also prevent sliding back of the stylet bundle. The rostrum is absent in sexuales of Pemphigidae and males of *Stomaphis.*

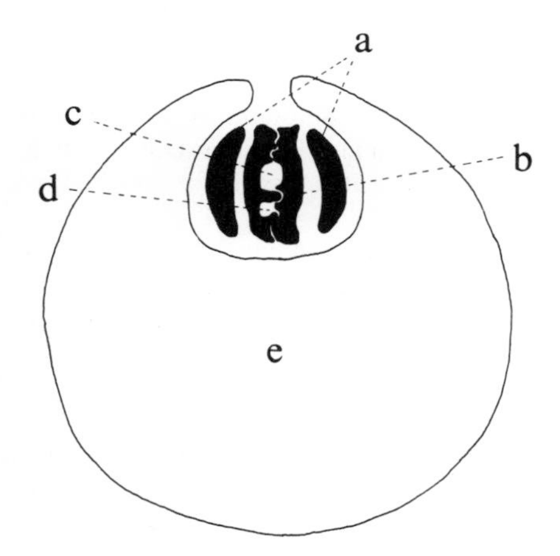

Fig. 19. Cross section of rostrum. – a = mandibles; b = laciniae of maxillae; c = food channel; d = saliva channel; e = labium.

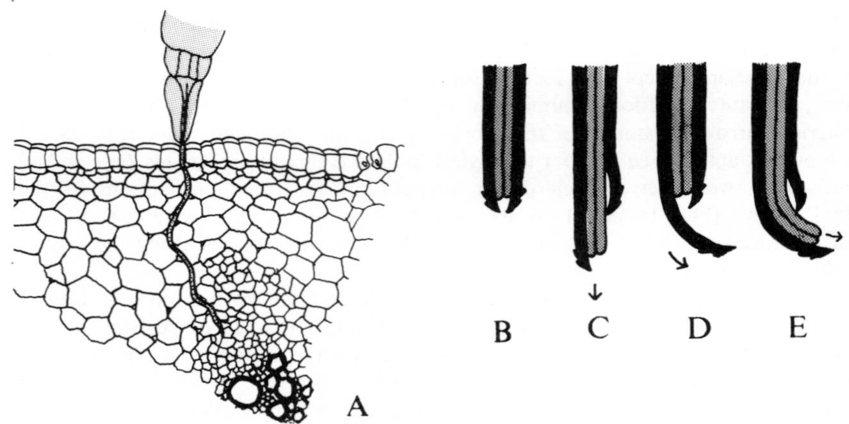

Fig. 20. The functioning of the stylets in the plant tissue. – A: the stylets are bored to the phloem between the cells. (Drawing based on photo in Nault). – B-E: apices of stylets; B: initial position; C: the maxillae are pushed forward together with a mandible; D: only one of the mandibles penetrates (followed by the maxillae); E (following D): the other mandible penetrates. (After Miles, redrawn). (C and D + E are alternatives).

Thorax

The thorax is better developed and more sclerotized in alatae than in apterae, especially mesothorax which contains the muscles moving the fore wings. The following description refers to alatae.

Prothorax is fairly weakly developed, forming a kind of »neck«. Pronotum is subdivided into an anterior and a posterior part by a transverse groove. Marginal tubercles occur on prothorax in many aphids. Prosternum is often reduced, consisting of small sclerites surrounded by membranous cuticle.

Mesotergum is vaulted and consists of notum and postnotum. Notum consists of a triangular praescutum, a large vaulted scutum, and a small, usually vaulted, scutellum. Scutum is subdivided into the mesothoracic lobes by a longitudinal groove. Its margins form part of the alary joints.

In repose the wings are either kept in a rooflike position, standing almost vertical, with the anterior edge turned down, or they are placed in a flat position, covering the abdomen, with the anterior edges turned outward (laterally). The latter position is found in Hormaphididae, Thelaxidae, and a few genera within other families. Aphids with wings rooflike in repose are more common. They have well developed mesothoracic lobes and an oblong praescutum (Fig. 21 A). Aphids with wings flat in repose have weakly developed mesothoracic lobes and a shorter praescutum (Fig. 21 B). In this case the longitudinal groove is hardly visible.

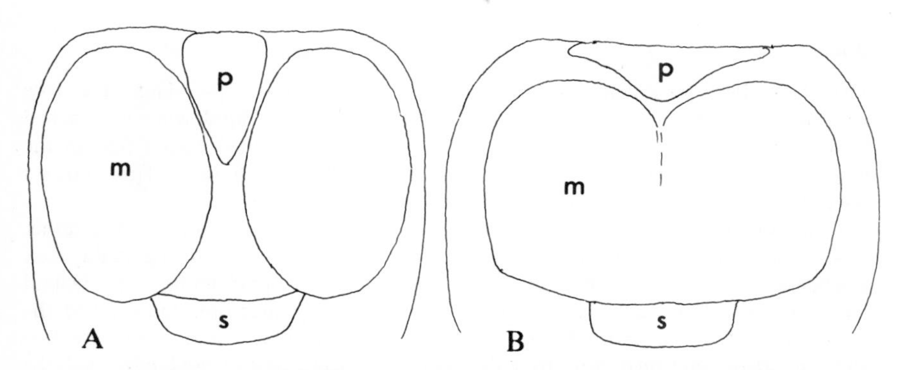

Fig. 21. Mesonotum of alate aphids. – A: species with wings in a rooflike position in repose; B: species with wings in a flat position: m = mesothoracic lobes (scutum); p = praescutum; s = scutellum.

Postnotum is small, often partly covered by scutellum.

Metathorax is reduced due to the small size of the hind wings, and because the movements of the hind wings during flight are governed by the movements of the fore wings. Metatergum is relatively small and narrow and never subdivided into notum and postnotum.

The thoracic segments of apterae are almost uniform and similar to the abdominal segments. Sclerotic articular forks (furca), which serve as support for the leg muscles, stretch from the sternites into the body. They are visible in mounted aphids made hyaline during preparation. Their shape is of importance to the taxonomy of some groups. The branches of the furca may be united or separated at the base (Fig. 22).

Fig. 22. Posterior part of thorax and anterior part of abdomen in ventral view, showing the meso- and metathoracic furca inside the body (visible in mounted specimens and here accentuated by black colour). – A: species with bases of mesothoracic branches united; B: species with branches separated.

Wings

The alatae have two pairs of hyaline wings with relatively few veins. The fore wing is much larger than the hind wing. During flight the hind wing is attached to the fore wing by 2–6 hooklets on its fore edge; the hooklets grasp a fold of the hind edge of the fore wing, the claval fold. Thus the mesothoracic flight muscles not only move the fore wings, but also the hind wings.

Venation of fore wing. Only two longitudinal veins are present, viz. the rather weak costa, running along the fore edge of the wing, and the strong main vein, which issues from the wing base to the well developed pterostigma, close behind the fore edge. The main vein is formed at the base by subcosta, radius, and the basal parts of media and cubitus. The distal parts of cubitus and media are free and run in oblique direction from the main vein towards the hind edge and the apex of the wing. Cubitus (Cu) has two branches, Cu_{1a} and Cu_{1b}, which are the two basal oblique veins, the latter closer to the wing base than the former. They leave the main vein, forming two separate oblique veins in most species. In a few others they have a common stem, and the free part of cubitus is forked. Media (M) usually has two forks, which is the primitive condition. In some aphids it has only one fork, or it is unbranched. The free part of media is usually invisible at the base. Radius (R) forms the posterior limit of pterostigma, and from this part of radius the radial sector branches. Radial sector (Rs) is not branched. It is sometimes weakly developed or absent.

Venation of hind wing. The only longitudinal vein, the main vein, is radius (or maybe the radial sector). Two oblique veins, corresponding to media and cubitus or to Cu_{1a} and Cu_{1b}, are usually present, but one or both may be absent.

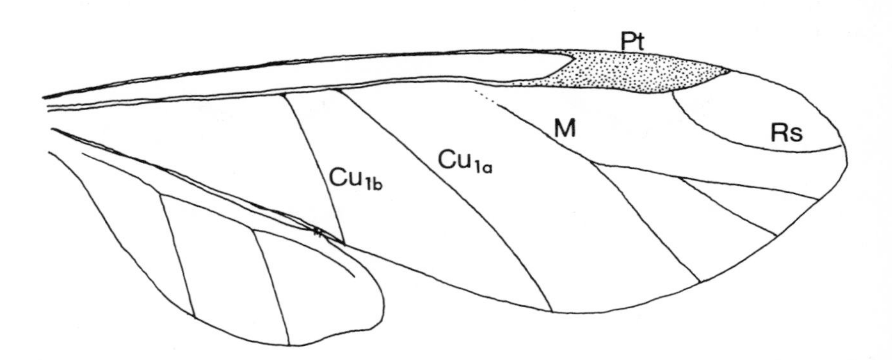

Fig. 23. Wings. – Cu_{1a} and Cu_{1b} = branches of cubitus; M = media; Pt = pterostigma; Rs = radial sector.

Legs

The fore coxae support the basal part of rostrum. The distance between fore and middle coxae is usually longer than the distance between middle and hind coxae. Some Phyllaphidinae have enlarged fore coxae whose muscles enable them to jump off by stretching the fore femora (Fig. 24 C). The jumping legs in aphids are the fore legs and not the hind legs, as in some other insect orders, which enable them to draw out the mouth parts from plant tissues before escaping. Trochanter is firmly connected with femur, sometimes fused with it.

Hind femur is longer and thicker than fore and middle femora except in species which jump by stretching the fore and middle tibiae (*Iziphya* and related genera) or the fore femora alone (*Drepanosiphum*) (Fig. 24 B).

Tibia is long and slender, with apical bristles or strong hairs preventing the aphid from sliding on rough surfaces. A sole bladder is well developed in aphids living on smooth leaves (Fig. 25). In oviparous females the hind tibia carries scent plaques which are small, roundish, angular or 8-shaped, pale fields (Fig. 3). The scent attracts the male. The oviparous morph only exceptionally lacks these organs.

Tarsus is usually two-segmented; it has only one segment in immature specimens and apterae of some Pemphigidae (Figs. 144 B, 246). Tarsi are absent in some

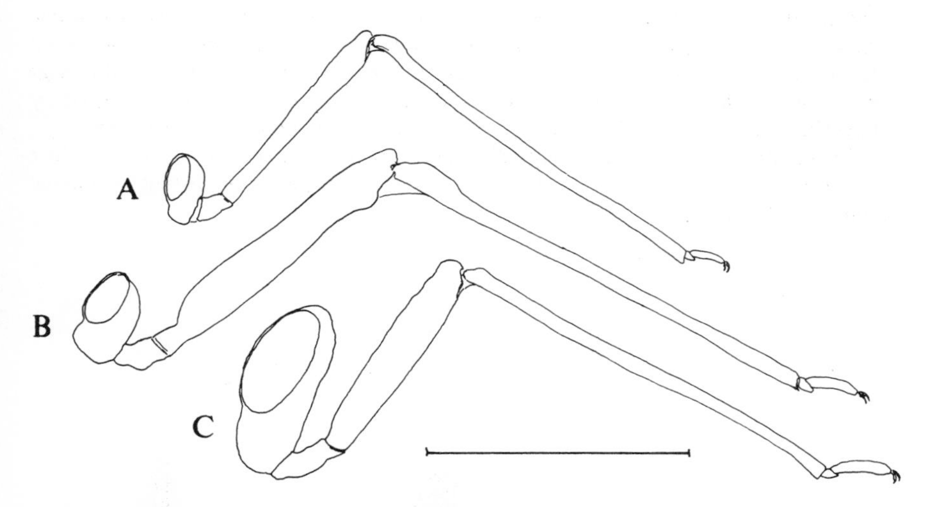

Fig. 24. Fore legs. – A: *Sitobion*, not jumping: B: *Drepanosiphum*, jumps by stretching out the tibia by means of strong muscles in the femur; C: *Tinocallis*, jumps by stretching out the femur by means of strong muscles in the coxa. (Scale 1.0 mm for A and B, 0.5 mm for C).

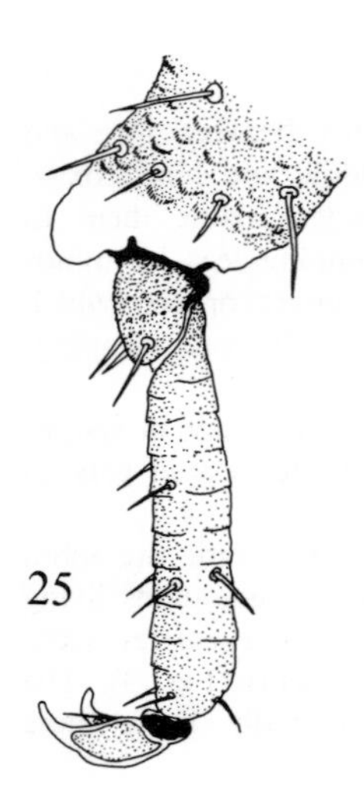

Fig. 25. Apex of hind tibia (with sole bladder) and hind tarsus of *Amphorophora ampullata* Buckt.

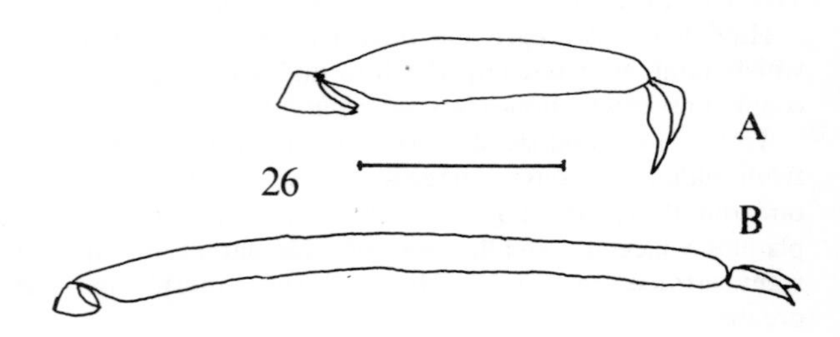

Fig. 26. Hind tarsus of A: *Phyllaphis fagi* (L.) and B: *Trama troglodytes* v. Heyd., the second segment being elongate. (A: scale 0.1 mm, B: scale 0.25 mm).

or all morphs of a few other aphids. The first segment is short, the dorsal side shorter than the ventral side. The number of hairs of this segment is a taxonomic character (Fig. 5 n); in descriptions the words »first tarsal segments with 3-3-2 hairs« mean that the number of such hairs is 3 on the fore legs, 3 on middle legs and 2 on hind legs. The second segment is longer, in hind legs of Traminae extremely long (Fig. 26 B). The length of this segment is often used as a diagnostic character.

There are two claws. The empodial hairs are situated below the claws; their length in relation to that of the claws and also their shape are important taxonomic characters (Fig. 27).

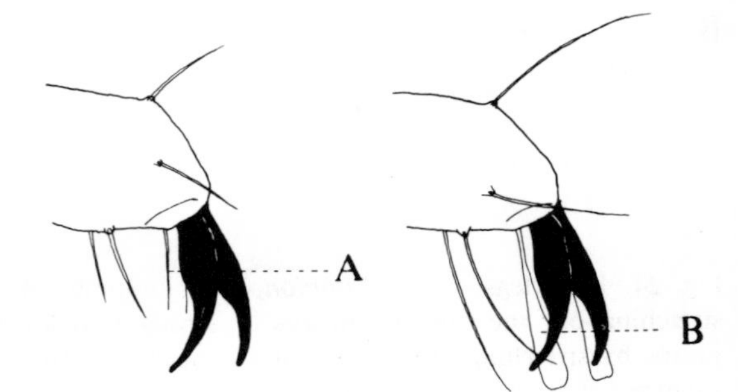

Fig. 27. Apex of second tarsal segment. – A: empodial hairs simple; B: empodial hairs spatulate.

Abdomen

Abdomen consists of nine visible segments, abdominal segments I-VIII and X, segment IX being reduced. The dorsal walls are called tergites. Cauda (the tail) is abdominal tergite X. Anus is situated below the cauda on the border between cauda and the ventral wall (the sternite) of abdominal segment X, called the anal (or subanal) plate (Fig. 28).

Abdomen is more strongly sclerotized and pigmented in alate females than in apterous females (and in males more strongly than in females) in many species, while the opposite is the case in others. Sclerotic and dusky or dark pigmented dorsal areas may be small or absent from some or all tergites. Small sclerotic areas surrounded by membranous cuticle are called sclerites or — if they are almost as small as hair bases — scleroites. They can be subdivided into segmental, marginal, pleural and spinal sclerites (Fig. 29) and intersegmental, usually pleural sclerites. The latter are more or less reticulate so-called muscle plates (Fig. 29 e). Sclerites in front of the siphuncular bases are called antesiphuncular sclerites (Fig. 29 f), and sclerites behind the siphuncular bases postsiphuncular sclerites (Fig. 29 g).

The dorsal sclerites may fuse into segmental pleurospinal cross-bands or cross bars. A large dorsal spot (or patch) may be formed by fusion of some of the cross bars, sometimes also of the marginal sclerites of the same segments. The mid-dorsum or the whole dorsum may then be entirely dark, or the dark area has pale »windows«, usually along the segment limits. Dorsal sclerites or short cross bars are often present on tergites (VI-) VII-VIII even if sclerites are absent from other tergites.

The anal plate and the sternite in front of it, the genital (or subgenital) plate, are usually pigmented. Spinal and marginal tubercles often occur, most frequently on segments I-VII.

Siphunculi

The siphunculi (or cornicles) are placed on abdominal segment V, usually at its

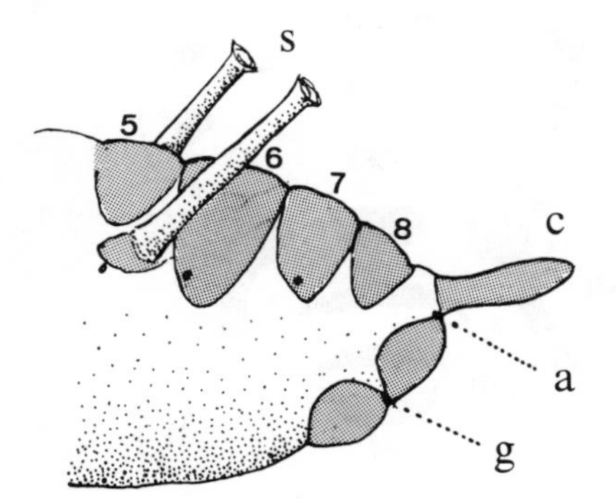

Fig. 28. Abdomen of viviparous female (Aphididae), left lateral view. - 5–8 = abd. segm. V–VIII; c = cauda; a = anus; g = genital aperture; s = siphunculus; the plate between a and g is the anal plate; the plate in front of g is the genital plate. (After Heie).

posterior margin (apparently situated between tergites V and VI) (Fig. 28 s). In some aphids they are placed on the anterior part of segment V, in a few others on segment VI.

The siphunculi are 0.3 × body length or shorter. Sometimes they only consist of low cuticular rings surrounding the siphuncular pores (ring- or pore-shaped siphunculi), and they may be so small that they are hardly visible (some Pemphigidae). Most aphids have siphunculi being 0.05–0.20 × body length. They may be straight, curved, or S-shaped. The shape may be cylindrical, swollen, or short and conical (stump-shaped). The aperture is usually situated at apex, in a few genera (*Aspidaphis, Aspidaphium*) on the side of the apical part. If situated at the apex the aperture is usually surrounded by a more or less distinct flange.

The siphuncular pores of Lachnidae are placed on strongly haired, conical warts, which are sometimes regarded as the siphunculi themselves (Fig. 30 K). Siphunculi seldom carry hairs, but often scales or rows of fine spinules making their surface look scabrous, imbricate, or reticulate. Reticulate siphunculi occur in species of only two groups of Scandinavian aphids, viz. Chaitophorini (Fig. 30 L) and Macrosiphini (Fig. 30 A).

A liquid waxlike secretion is produced by abdominal glands at the siphuncular bases and released through the apertures. A spot differing in colour from the rest of the body can be seen at the base of either siphunculus in live specimens of some species with membranous (somewhat hyaline) abdomen, e.g. in *Aulacorthum solani* and *Rhopalosiphum padi.*

When not secreting the aphid closes the siphuncular aperture with a cover attached to the inner wall just below the aperture. The cover is pulled down by a longitudinal muscle when secretion is produced.

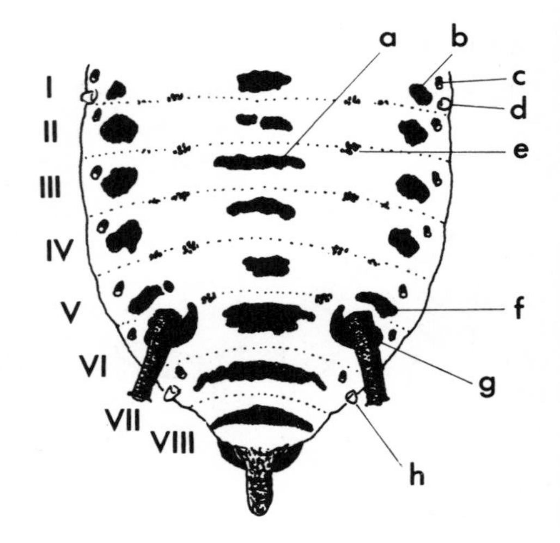

Fig. 29. Dorsal pigmentation of abdomen of alate viviparous female of *Aphis fabae* (Scop.), hairs omitted. – I–VIII = abd. segm. I–VIII; a = spinal sclerite; b = marginal sclerite; c = stigmal pore placed at the posterior margin of a stigmal sclerite; d = marginal tubercle on abd. segm. I; e = intersegmental pleural muscle sclerite; f = antesiphuncular sclerite; g = postsiphuncular sclerite (behind the base of the siphunculus); h = marginal tubercle on abd. segm. VII.

The secretion consists of a liquid with large, strongly light-refracting, globular cells deriving from the fat tissue at the siphuncular base and from the haemolymph. It appears as droplets from one or both siphunculi when the aphid is alarmed by an unusual stimulus, e.g. tactile contact with a pointed object. Previous to that the siphunculus is raised by muscles. The secretion contains a volatile substance, $C_{15}H_{24}$, which acts as an alarm pheromone.

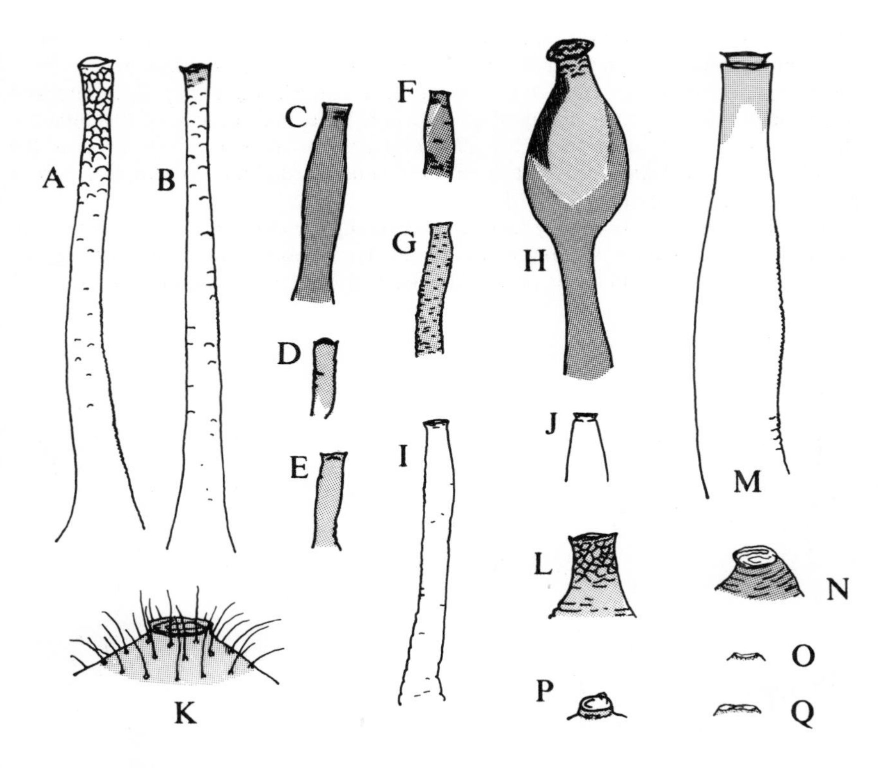

Fig. 30. Siphunculi. – Aphididae: A: *Macrosiphum euphorbiae* (Thomas); B: *Acyrthosiphon pisum* (Harris); C: *Megoura viciae* Buckt.; D: *Hyalopterus pruni* (Geoffr.); E: *Hayhurstia atriplicis* (L.); F: *Brevicoryne brassicae* (L.); G: *Lipaphis erysimi* (Kalt.); H: *Rhopalosiphoninus calthae* (Koch); I: *Myzus persicae* (Sulz.); J: *Brachycaudus helichrysi* (Kalt.). – Lachnidae: K: *Cinara pilicornis* (Hartig). – Drepanosiphidae: L: *Periphyllus testudinaceus* (Fern.); M: *Drepanosiphum platanoidis* (Schr.); N: *Callaphis juglandis* (Goeze); O: *Phyllaphis fagi* (L.) – Thelaxidae: P: *Thelaxes dryophila* (Schr.). – Pemphigidae: Q: *Eriosoma ulmi* (L.). (After Heie).

Siphunculi are absent from some aphid species. In some species they are present in some generations and absent from others (e.g. *Thecabius affinis* and *Longicaudus trirhodus*), or they may be present in some individuals in a single generation and absent from others (e.g. the fundatrix-generation of *Colopha compressa* and the alate spring generation of *Tetraneura ulmi*). The shape of the siphunculi may differ in the different generations of a species.

Cauda

The cauda (the tail) is short and rounded, little-developed and similar to the anterior abdominal tergites in some aphids, and well developed, projecting more or less above the anus, and then semicircular, triangular, tongue-shaped, finger-shaped, or knobbed in the adult stage of other aphids (Fig. 31). It is never well developed in immature stages. Cauda may be more or less distinctly constricted at the base or near the middle. It is called knobbed if it is strongly constricted, and if the apical part is globular or nearly globular (Fig. 31 E).

Aphids visited by ants usually have a short or slightly developed cauda. The main function of the cauda is to prevent the liquid, sticky excrements from flowing out over the body. The aphid will raise abdomen and bend its cauda upward when an

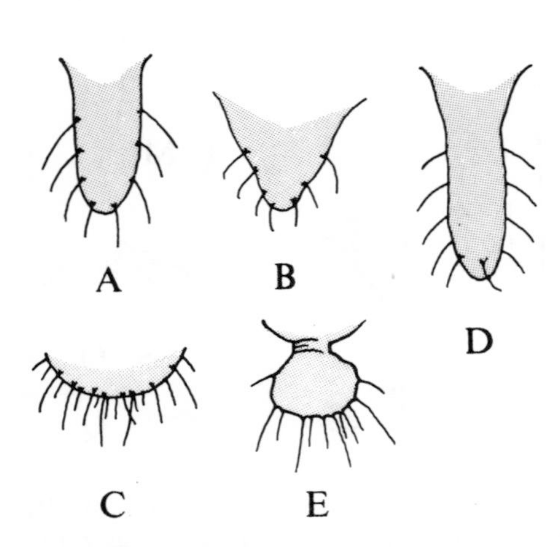

Fig. 31. Shape of cauda. – A: tongue-shaped, *Aphis*; B: triangular, *Brevicoryne;* C: broadly rounded or semicircular, *Cinara;* D: finger-shaped, *Macrosiphum;* E: knobbed, *Tuberculatus*.

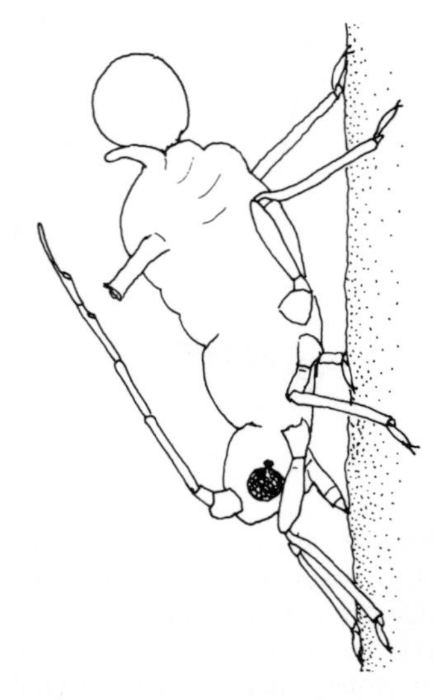

Fig. 32. Feeding aphid producing an excrement droplet. (After Kunkel, redrawn).

excrement droplet appears from the anus (Fig. 32). The usual position of aphids feeding on stems and other parts of the host plant is head down, so that the cauda prevents the droplet from rolling down along the dorsum before it is thrown away by contraction of the rectal muscles or by a kick of a hind leg. In aphids attended by ants the droplet is removed by the ants.

In aphids with a strong wax production (e.g. most species of Pemphigidae) the body and often the excrement droplets are covered by hydrophobe wax threads or powder and a well developed cauda therefore is not necessary to remove the sticky substance.

External genital organs

The female genital aperture is a transverse ventral cleft (Fig. 33 A, b) between the anal plate (Fig. 33 A, d) and the genital plate (Fig. 33 A, a). Abdominal sternite IX is represented by the so-called rudimentary gonapophyses, which are small,

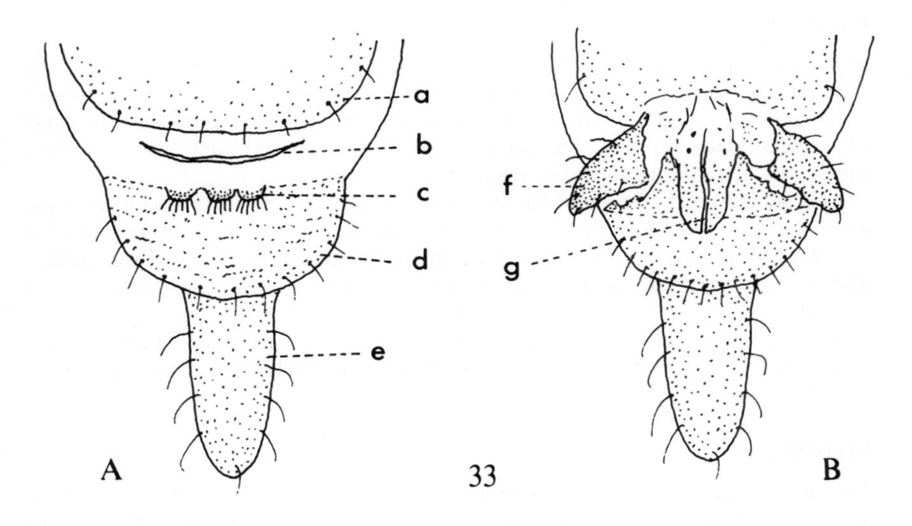

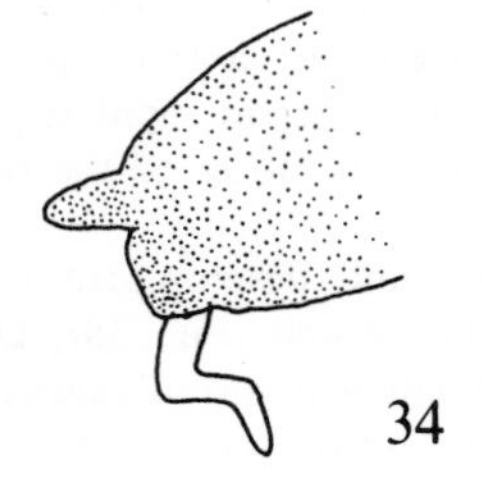

Fig. 33. Posterior part of abdomen of A: female and B: male of the family Aphididae. – a = genital plate; b = female genital aperture; c = rudimentary gonapophyses; d = anal plate; e = cauda; f = claspers; g = basal part of penis. (After Szelegiewicz, redrawn).

Fig. 34. Posterior part of male abdomen of *Myzus,* with the soft distiphallus projecting.

usually densely haired, processes at the anterior margin of the anal plate (Fig. 33 A, c). The hairs are called gonochaetae.

The male genital apparatus is relatively large, and consists of sclerotic parts, wherefore it is easy to determine the sex of an aphid, even with the naked eye. Aedeagus is situated ventrally on abdominal segment IX. Only the sclerotic basal part is visible in repose (and in mounted specimens) (Fig. 33 B, g). During coitus the distal soft S-shaped part is extended by an inner liquid pressure (Fig. 34). Two short, finger-shaped, conical or leaf-shaped parameres (or claspers) (Fig. 33 B, f) are situated laterally of the aedeagus.

The male genitalia have so far not been used in aphid taxonomy, also because males are relatively rare.

Bionomics and ecology

Life cycles

In most aphids fundatrix emerges from a fertilized, overwintering egg in spring, and the fundatrix generation is succeeded by several generations of parthenogenetic viviparous females. The sequence of generations ends in the autumn with appearance of sexuales, males and oviparous females. The females lay fertilized eggs from which fundatrices hatch during the following spring.

This annual cycle of generations is called a holocycle, and aphids with this cycle are holocyclic (Fig. 35). Aphids are called anholocyclic if they do not produce sexuales. All generations of anholocyclic species consist of parthenogenetic females. They do not overwinter as eggs, but as adult or immature parthenogenetic females (Fig. 36).

Some species are holocyclic in temperate climates, and anholocyclic in the tropics, and in Scandinavia some species have both kinds of cycles and methods of overwintering.

Reproduction

The unfertilized eggs of the viviparous females cleave rapidly; the first embryos start development before the birth of their mother so they are present in their grandmother. The first nymph is born as soon as the mother has moulted for the last time. The posterior end appears first, and the nymph is covered by a thin membrane which bursts during birth (Fig. 37). The births follow each other in rapid sequence, usually 2–5 births per day, depending on the morph and generation of the mother and some environmental factors.

The eggs of the oviparous females need fertilization, and development is late. Copulation is short. The male mounts the female from behind (Fig. 38). During copulation the female keeps quite still. One male may copulate with several females.

The eggs are laid more or less concealed in the vegetation above or below the

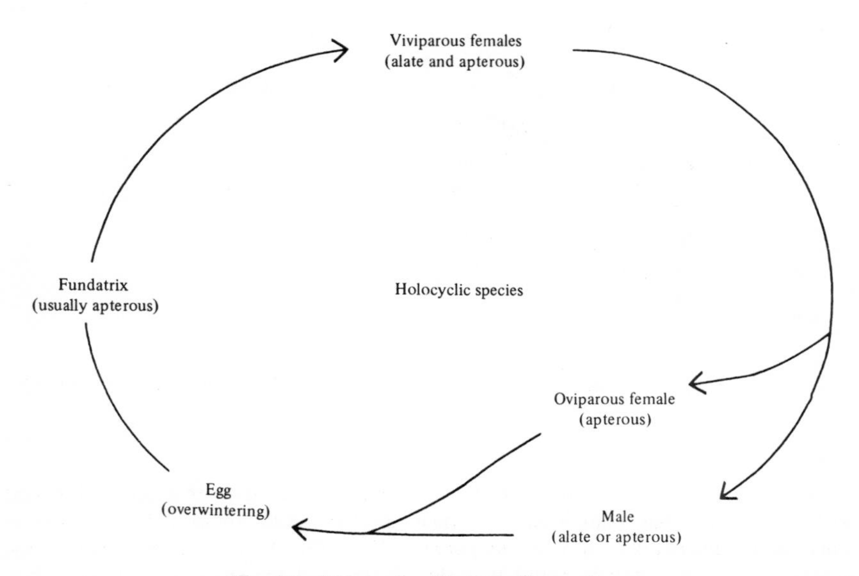

Fig. 35. Life cycle of holocyclic species.

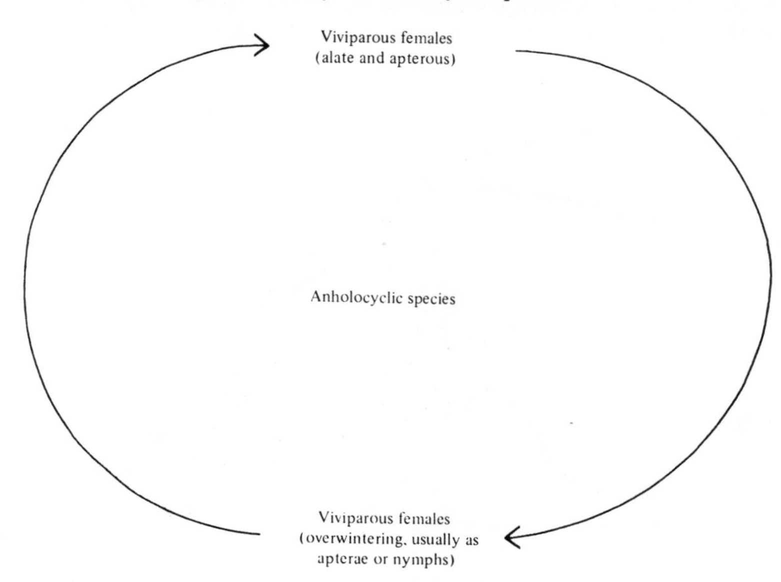

Fig. 36. Life cycle of anholocyclic species.

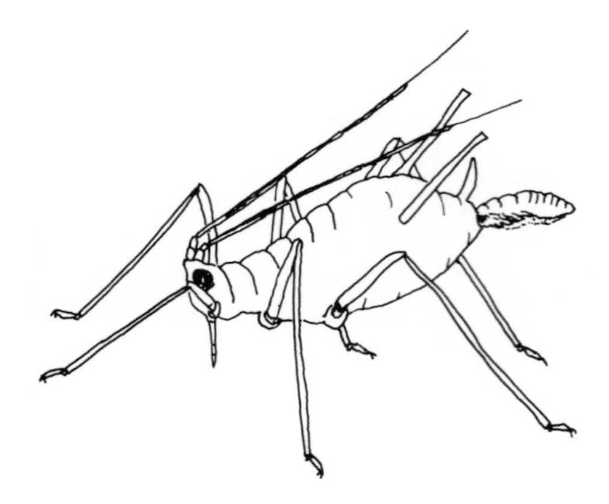

Fig. 37. *Uroleucon* giving birth.

ground, e.g. in crevices in the bark. On trees and bushes they are often laid in bud axils. They are big compared with their mothers, e.g. about 0.7 mm long in *Euceraphis punctipennis* and *Myzus persicae* (Fig. 39), whose eggs have about the average size. The width is about half the length, and the shape oval (ellipsoidic). The colour is yellow or green just after the egg has been laid, turning black after a few days if fertilized, shining black in most genera, but dull in e.g. *Mindarus*, whose

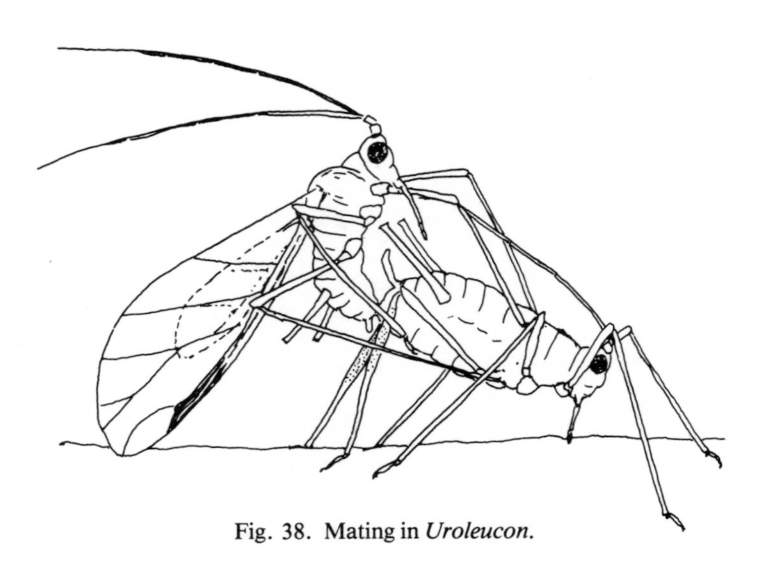

Fig. 38. Mating in *Uroleucon*.

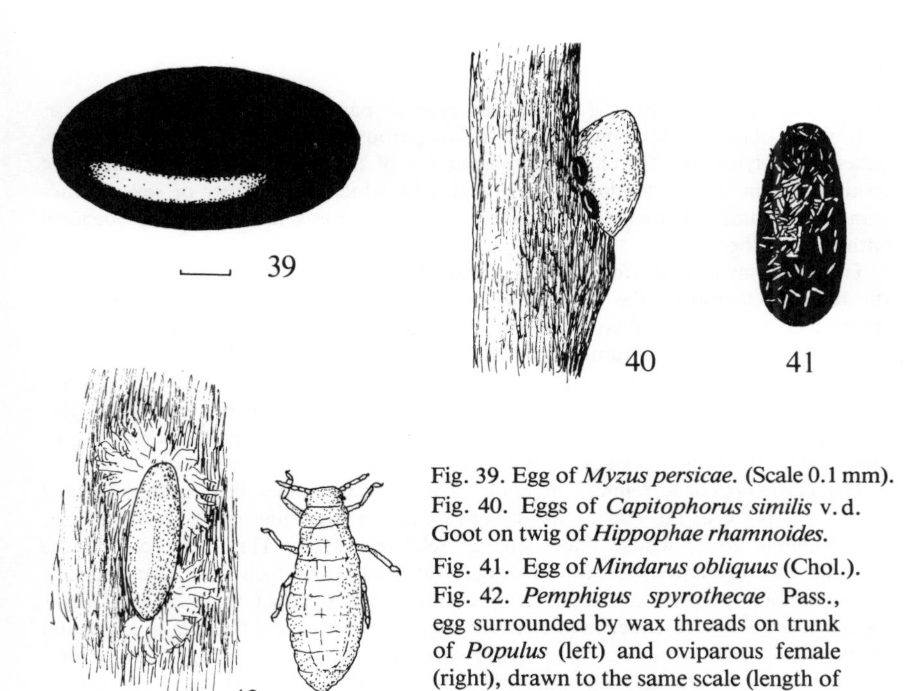

Fig. 39. Egg of *Myzus persicae.* (Scale 0.1 mm).
Fig. 40. Eggs of *Capitophorus similis* v.d. Goot on twig of *Hippophae rhamnoides.*
Fig. 41. Egg of *Mindarus obliquus* (Chol.).
Fig. 42. *Pemphigus spyrothecae* Pass., egg surrounded by wax threads on trunk of *Populus* (left) and oviparous female (right), drawn to the same scale (length of oviparous female 0.8 mm).

eggs are wax powdered (Fig. 41). A secretion from glands inside the female genital duct covers the egg. It hardens shortly after, gluing the egg on to the plant surface. In most Scandinavian species the eggs are laid in October, November, and December. Some of the oviparous females avoid falling to the ground with the foliage in the autumn because they move to the branches or the trunk.

The sexuales of Pemphigidae are dwarfish and without rostrum and are usually born on trunks. They do not grow during their larval instars. The oviparous female lays only one egg, which is almost as long as itself (Fig. 42), and soon after she dies. Most other aphids lay more than one egg, in Aphididae 4–8 per oviparous female, in Thelaxidae 2.

The egg hatches in spring. A warm period in the middle of the winter does not make the young fundatrix appear, because the egg is in diapause. In some Scandinavian aphids the diapause lasts until February. In Denmark the eggs normally hatch in April, in years with high temperatures as early as in the middle of March, in cold years as late as the beginning of May.

Morph determination

All eggs are diploid, also the unfertilized eggs of parthenogenetic females. Meiosis only takes place in the genital organs of oviparous females and males. Parthenogenesis is thelytokous. The chromosome number of the parthenogenetic ovum is not reduced except in ova developing into males, in which the number of sex chromosomes is reduced. Females produced by parthenogenesis are genetically identical with their mother.

The male has one chromosome less than the female, XO compared with XX in the female. Obviously, the sex of the individual reproduced by parthenogenesis is determined before determination of any other morph alternatives happens. Generally, males occur only in the autumn. Environmental factors at that time affect the parthenogenetic females in such a way that all or some of them reproduce males. Not only the maternal, but also the grandmaternal environment can be decisive, at least in species whose males are born by specialized viviparous females, sexuparae or androparae, since the first embryos start development before their mother is born.

The parthenogenetic females of many Scandinavian species bear both male and female offspring at the end of the summer and the beginning of the autumn, e.g. *Macrosiphum lisae*, whose males appear relatively early. They are born in the last part of July and first part of August by females, which also bear females. These mothers produce about 25 nymphs, about one-third of these males. In this species there is first a sequence of 8–10 females, then males and females in accidental order (Heie 1965). The fact that male-bearing mothers start to reproduce several females seems to indicate that the male determination is maternal and not grandmaternal in this species.

The apterous morph of *Megoura viciae,* investigated by Lees (1959, 1961, 1966), is genetically disposed to bear 11% males and 89% females (100% = 90 nymphs); sometimes less than 11% males, never more, and these are born in the middle of the reproductive period of their mother. Various influences prevent birth of males in the greater part of the year. The maximum number (11%) appeared when the mothers grew up at a temperature of about 15°C, while few, or no males, were born at temperatures below or above 15°C.

In other species the production of males is repressed when days are long during the development of their mothers. When the day length has decreased to 12–13 hours in September, most Scandinavian aphid species will produce males.

High temperatures or long days or both prevent males from being born in summer. The restraint on the formation of males weakens at the end of the summer due to falling temperature and shorter days, and the parthenogenetic females then begin to bear males. Though temperature is low in spring and the days short, males do not appear that early, because male production is prevented by an inner mechanism. This mechanism works in a certain period, not in a certain number of generations, and it may depend on a substance produced in the fundatrix (or in the fertilized egg), multiplied step by step with cell division, and gradually losing its effect.

The egg cell develops into a female if it has two X-chromosomes. The fertilized egg always gives rise to a female (fundatrix) because it receives an X-chromosome from both parents. Spermatocytes without an X-chromosome degenerate, so all sperm cells carry an X-chromosome. The unfertilized eggs in the ovarioles of parthenogenetic viviparous females also have the full diploid complement of chromosomes, including the X-chromosomes, when male production is prevented by some of the above-mentioned factors. The next alternatives are: sexualis-female (ovipara) or parthenogenetic female (vivipara).

Oviparae appear during the autumn in most Scandinavian aphids, usually a little later than the males. Short-day conditions (or rather long-night conditions) may provoke oviparae (and also males) to appear at other times of the year, at least in some species.

The embryos are primarily determined as oviparae, but during long days the mother produces a hormone which changes the development so they become viviparae.

The mechanisms must be different in different species. Sexuales of arctic aphids are produced under long-day conditions, and they appear shortly after the fundatrix-generations. Something like that happens in *Mindarus abietinus, Glyphina betulae, Aphis farinosa,* and *Cinara pilicornis* on lower latitudes. In these (and other) species sexuales are born in summer. *Mindarus abietinus* has only three generations a year, the fundatrices, their alate offspring (parthenogenetic females) and the sexuales. Obviously inheritance is stronger than the environmental factors in this case.

The appearance of oviparae of *Megoura viciae* is determined by day length and temperature. An inner mechanism similar to the mechanism regulating the male production prevents their appearance in spring. Repression of the ovipara production in spring in *Megoura* is ascribed to an effect of a substance, which is present in the fundatrix and passed on to her offspring. The effect is lost after 25–45 days at 20° C, independent on the number of generations. After that time the apterous mother bears oviparae at 15°C, provided the day length is 14^h55^m or shorter. The critical day length becomes three minutes shorter if the temperature rises by one degree. The days need not be very short at low temperatures for oviparae to appear. The higher the temperature, the shorter the day must be. Oviparae of *Megoura viciae* are not produced at temperatures above 21°C, no matter how short the day may be.

Many root-feeding aphids produce sexuales or sexuparae in the autumn although they live in permanent darkness. The day length may work indirectly through the host plant in some cases.

The parthenogenetic female embryo or young nymph has two alternative courses of development to the adult stage. She may become alate or apterous. The alate viviparous female should be regarded as the more primitive or original viviparous morph, althoug the apterous female usually is the more common morph, and accordingly treated in greater detail in descriptions and keys. The apterous viviparous female is a reduced morph, a specialist on reproduction, while the alate morph provides for dispersal of the species.

The number of births per apterous female is not large compared with many other animals, but it is larger than in an alate female. Due to parthenogenesis and rapid growth of embryos the ability to multiply is huge. The generations succeed each other very quickly. In addition the young are able to care for themselves from birth.

The number of instars is usually four, in apterae as well as in alatae, but the stages last for a shorter time in apterae than in alatae. The aptera starts reproduction immediately after the last moult. It usually does not run the risk of unsafe roaming and loss of energy, and it does not use substances and energy for development of wings, wing muscles, and some sense organs. The surplus of food and energy is used for development of embryos.

If neoteny means retention of certain larval characters at the mature stage, then apterous aphids are neotenic, but they are not neotenic if neoteny means acquisition of sexual maturity by some larval instar. The apterous adult is a proper adult, an alternative of the alate adult, with typical adult characters in excess of the sexual maturity and ability to reproduce, e.g. the typical adult shape of cauda, though it resembles the immature aphids in several respects.

Finding the factors determining the choice between alate and apterous morphs often meant finding those promoting wing development, although it was more important to find the factors leading to winglessness, since the alate morph should be regarded as the primary adult morph and the apterous morph as the secondary. Production of juvenile hormone until a much later stage of development of the aptera than of the alata may be the answer or part of the answer (Blackman 1974).

Several investigations have shown that in many species isolation resulted in appearance of apterae, and crowding (or contact with conspecific specimens) appearance of alatae, but also that varying response to environments exists within aphids. All viviparous females of *Drepanosiphum platanoidis* and some other species are alate, while nearly all are apterous in *Phloeomyzus* spp. (not known in Scandinavia). Response to various factors differs in the different species, sometimes in different populations of the same species.

Bonnemaison (1951) found that several nymphs became alatae in *Brevicoryne brassicae,* if they grew up close to the mother, contacting her for at least 72 hours (at 20°C). If they did not touch their mother, or if contact ceased before they were 72 hours old, they all became apterous. The effect of this factor is not only postnatal, but also prenatal. If the mother had no contact with other aphids of the same species since her third instar, none of her nymphs born during the first three days after her last moult became alatae even if they were in constant touch with her.

In *Myzus persicae* the effect of the same factor is exclusively postnatal. It is the contact with other nymphs, not the contact with the mother or the mother's contact with other individuals, which determines the morph in this species. Isolated nymphs of *M. persicae* nearly always become apterous, and crowded nymphs usually become alate.

The effect of this factor is prenatal in *Megoura viciae* (Lees 1967). Isolation or crowding of mothers are decisive in this species. Isolated mothers bear nymphs developing into apterae, and crowded mothers nymphs developing into alatae.

Starvation or lack of proper food items are not the basic causes for the crowding effect in this species, although the advantage for the aphid colony is the chance to avoid starvation through transfer from an over-populated to a fresh plant. The isolation of mothers in *Megoura* results in production of apterae even if the mothers do not feed. Poor food supply may however have indirect effect because it involves restlessness on a wilting host plant so that the parthenogenetic females will often touch each other.

Apperance of alatae in other species is a direct result of insufficient food, even if crowding does not occur, e.g. in *Acyrthosiphon pisum* (Sutherland 1969).

Apart from »isolation-or-contact« and poor food supply many factors are important, e.g. temperature, day length, ants, and parasitism. A connection between the mother's morph and that of the daughter apparently also exists. Among the nymphs born by apterous mothers a greater percentage usually becomes alate than among nymphs born by alate mothers.

Host plant relationships

Most aphids feed on sieve tube sap, which consists of an aqueous solution of sugar (sucrose) and other substances, primarily organic substances, among which the amino acids are particularly important to aphids. A few aphids feed on cells in parenchymatic tissues.

Some aphids are monophagous, feeding on one kind of plants, some are oligophagous, feeding on a few plant species, and some are polyphagous, feeding on many plants belonging to different families. Some species within Anoeciidae, Pemphigidae, and Aphididae have host-alternation. They are called heteroecious or dioecious, while aphids without host-alternation are called monoecious. Dioecious aphids go from primary hosts, which are woody plants, trees or bushes, to secondary hosts, which usually are herbs, and back again. The fundatrices feed on the primary hosts. Alate parthenogenetic females of the second generation (and/or the succeeding ones) fly, usually in spring and early summer, to the secondary hosts, where several parthenogenetic generations develop. Aphids flying from the secondary to the primary hosts in the autumn are either sexuparae (Anoeciidae and Pemphigidae) (Fig. 43) or gynoparae and males (Aphididae) (Fig. 44).

Some dioecious species are able to accomplish their cycle solely on the primary host, e.g. *Macrosiphum rosae* on rose. Their migration to secondary hosts in summer is called facultative, while other dioecious species, needing secondary hosts to complete their life cycles, have obligatory host-alternation. The border between facultative and obligatory host-alternation is however indistinct. *Hyalopterus pruni* and a few other species can stay on their primary hosts throughout the summer, but the males of these all derive from colonies on secondary hosts, so the accomplishment of their life cycles depends on host-alternation of part of the population.

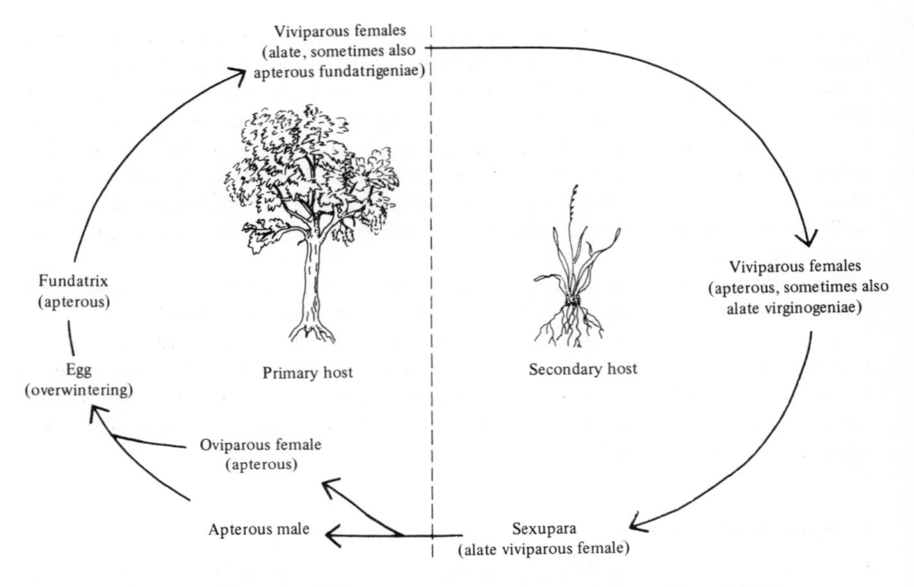

Fig. 43. Life cycle of dioecious species of Anoeciidae or Pemphigidae.

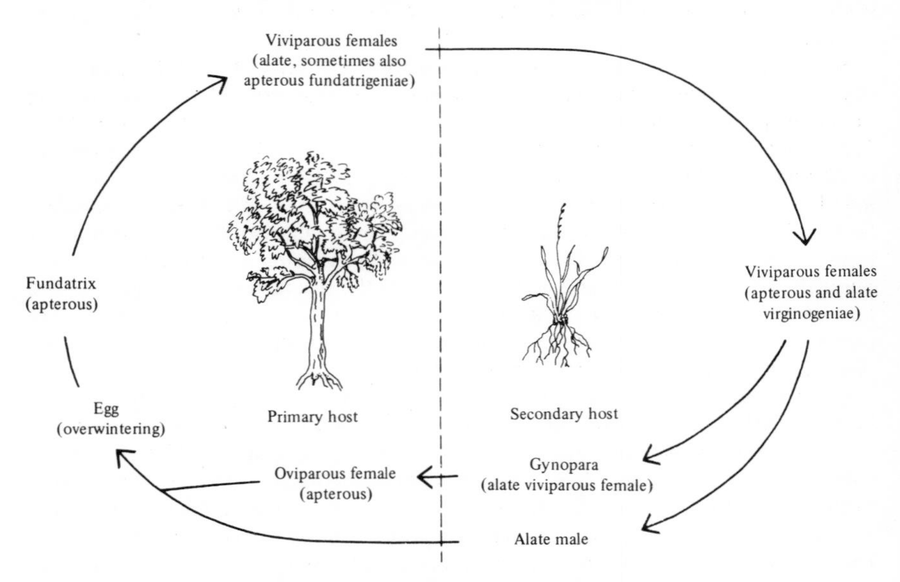

Fig. 44. Life cycle of species of Aphididae with obligatory host-alternation.

Host-alternation may be considered an adaptation to nutritional conditions of the two kinds of hosts, woody plants and herbs. The sieve tube sap of woody plants is richer in nitrogen compounds in spring and autumn than in summer (Mittler 1958, Auclair 1963). Their leaves are of nearly equal age, and their proteins are built up in spring and decomposed in autumn. In summer the sieve tubes are used almost exclusively to transport sugar from the leaves to the rest of the plant. In this period the amino acid contents, and also the nutritional value, are minimal. The leaves of herbs — to the contrary — are of different age. That means that aphids can find places with optimal nutritional conditions on herbaceous plants during the summer months, places where proteins are built up, viz. young leaves, upper parts of stems, and roots, and places where proteins are decomposed, viz. old, low-sitting leaves. Roots of woody plants may also be suitable food sources in summer.

New shoots with young leaves are produced in summer by woody plants which have been defoliated (by insects or artificially, e.g. in hedges). Aphids using these plants as primary hosts occasionally survive and reproduce on such new shoots for a longer period than on the normal shoots developed in spring.

Host-alternation is not necessary to all aphids living on woody plants. Many species are monoecious on trees or bushes. Some have a shortened life cycle and develop sexuales in early summer, e.g. *Aphis farinosa* on *Salix*. Some species of *Periphyllus* survive on *Acer* as special diapause nymphs in the summer (''summer larvae'') with prolonged first instar. Other monoecious aphids on trees are not specially adapted to the decrease of nitrogen in the sieve tube sap in summer, e.g. several species of Drepanosiphidae. They are more numerous at the beginning of the summer and in the autumn than in midsummer. Reproduction may then be very low.

Most aphids are monoecious on herbs. They mainly feed on upper parts of stems, young leaves, old leaves, or on subterranean stems and roots, rarely on fresh, green, mature leaves. They are often mimetic, with the same colours and colour pattern as the preferred parts of their host plants, an evidence of the steadfastness of their feeding preferences.

The water balance of plants also affects the choice of feeding places. Plants and plant parts with a high turgor pressure are often preferred. Aphids respond to decreasing velocity of sap inflow through the stylet channel by withdrawing the stylets and moving to another place. The response may be different in different species. *Myzus persicae* and *Acyrthosiphon pisum* gather on terminal shoots of plants in danger of exsiccation, while in such cases *Aphis fabae* spread all over the plant. Changes in turgor pressure are less important to some species, e.g. *Myzus persicae,* than to others. On succulent kale plants alatae of *M. persicae* land on old leaves as well as on terminal shoots and young leaves, while on exsiccated kale plants they primarily land on terminal shoots, preferring these to shoots of succulent plants. The behaviour of *Brevicoryne brassicae* is different. On succulent kale the alatae of this species primarily settle on mature leaves in the middle of the plant as well as on the terminal shoots, to a lesser degree on young and old leaves, and exsiccated kale plants hardly have any attraction to *Brevicoryne* (Wearing 1972).

Some aphids move from one part of the host to another during the year. *Nasonovia pilosellae* on *Hieracium* feeds on rosette leaves close to the ground in early summer and on upper parts of stems and inflorescences in late summer.

The choice of host is influenced not only by physiological conditions of the plants and the individual parts of the plants, but also by other properties of the hosts. Most aphids are fastidious as to a plant species, especially monophagous aphids, and some of the oligophagous species are »competent botanists« choosing closely related host plants.

It is difficult to imagine that a species is so specialized that its right composition of nutrition only occurs in a single or a few plant species. The explanation of the specific choice of hosts may be found in the heritable behaviour of aphids rather than in the specific physiology of plants. This is concluded from the fact that each of the species occurring together on a definite host plant often has other hosts, which are not hosts of the others. Plant-specific chemical compounds devoid of nutritional value, but acting as taste stimuli to certain aphid species, may in some cases explain host preferences. *Brevicoryne brassicae* feeds on cabbage, kale, and other Cruciferae, and also on *Tropaeolum,* which is not allied with Cruciferae. It has been proved that *Brevicoryne* is attracted by sinigrin, which occurs both in Cruciferae and in *Tropaeolum.*

The oligophagous *Brevicoryne brassicae* and the polyphagous *Myzus persicae* both feed on *Brassica,* but while *M. persicae* is attracted by the nutritional value of the sap, especially its contents of amino acids, is *B. brassicae* attracted by the sinigrin. *M. persicae* prefers young and old leaves to mature leaves, but *B. brassicae* prefers young and mature leaves to old ones, in spite of the greater nutritional value of the old leaves, because it gets more sinigrin the greater the turgor pressure is.

The signals to which the aphid responds when searching for a host plant and suitable feeding places on it are mainly taste stimuli. The colour of the plant, tactile stimuli from hairs or wax covering of leaves and stems, and probably also olfactory stimuli, may, however, also be of importance.

The advantage of being monophagous is increased possibility of specific adaptation to the host plant, consequently also acquisition of increased superiority in the competition with other less adapted species.

In monoecious species the hosts usually belong to one particular genus or family. The primary and secondary hosts of a dioecious aphid are, however, never closely related. The primary hosts of a dioecious aphid with several host plants are mutually related, and its secondary hosts are usually also mutually related, but the secondary hosts are not relatives of the primary hosts. The primary hosts of *Rhopalosiphum insertum* are thus members of Pomaceae, while the secondary hosts are members of Gramineae.

Species of the same genus often have mutually related hosts. The species of

Pleotrichophorus, which are monoecious, all feed on Anthemideae within Compositae. Dioecious species belonging to the same genus usually have interrelated primary hosts, sometimes also interrelated secondary hosts. Most species of *Cavariella* are thus host-alternating between Salicaceae (I) and Umbelliferae or Araliaceae (II). In genera comprising both dioecious and monoecious species the latter often feed on plants related to the secondary hosts of the former. *Metopolophium dirhodum* migrates between *Rosa* (I) and grasses (II), while the other species of *Metopolophium* are monoecious on grasses. In such cases the monoecious species obviously evolved from dioecious ancestors.

In particular within Pemphigidae the host spectra of most genera show that the association with the primary hosts is older than that with the secondary hosts. The primary hosts are also here more closely allied with each other than are the secondary hosts, and, moreover, special adaptations evolved in connection with the life on the primary hosts, e.g. formation of galls. The species of *Eriosoma* and related genera use *Ulmus* as primary host, and species of *Pemphigus* and some related genera use *Populus,* while the secondary hosts of the dioecious members of both systematic subgroups of Pemphigidae belong to many plant families. It is the opposite in *Prociphilus* (Pemphiginae). Its primary hosts belong to different families of deciduous trees (Oleaceae, Caprifoliaceae, Pomaceae), while the secondary hosts of nearly all species are conifers. This means that the connection with the primary hosts in this case happened fairly late, after establishment on the plants, which are the secondary hosts today (Hille Ris Lambers 1950a).

In accordance with the Fahrenholz rule that interrelated organisms have interrelated parasites, aphids often display interrelationships of plants, but far from always. Several examples show that acquisition of new hosts occurred several times during the geological history of the aphids.

Alate aphids do not only land on well established hosts, but on many others, more or less accidentally. They usually leave again after a while, probably repelled by the taste of the sap. If not repelled, for instance due to change of the genetical basis for taste response by mutation or recombination of genes, they may in rare situations colonize the ''new'' plant. Host specificity may also depend on environmental factors. It tends to break down in some aphids in arctic climates. *Acyrthosiphon auctum* and some other species have been found on more plant species in Iceland, than in Denmark, although these extra hosts are not restricted to Iceland, but are common plants in both countries.

Aphidologists will have observed aphid colonies on unusual food sources several times. Firm connections seldom develop between these aphid species and previous non-hosts, probably because new combinations of genes are formed once a year when the sexuales produce the fertilized eggs. The genotype preadapted for feeding on a different plant species might be highly multiplied through diploid parthenogenesis in one summer, but may have little or no chance to reappear next year, unless the aphid is — or becomes — anholocyclic.

Adaptation to the host plants is morphological as well as physiological. Aphids

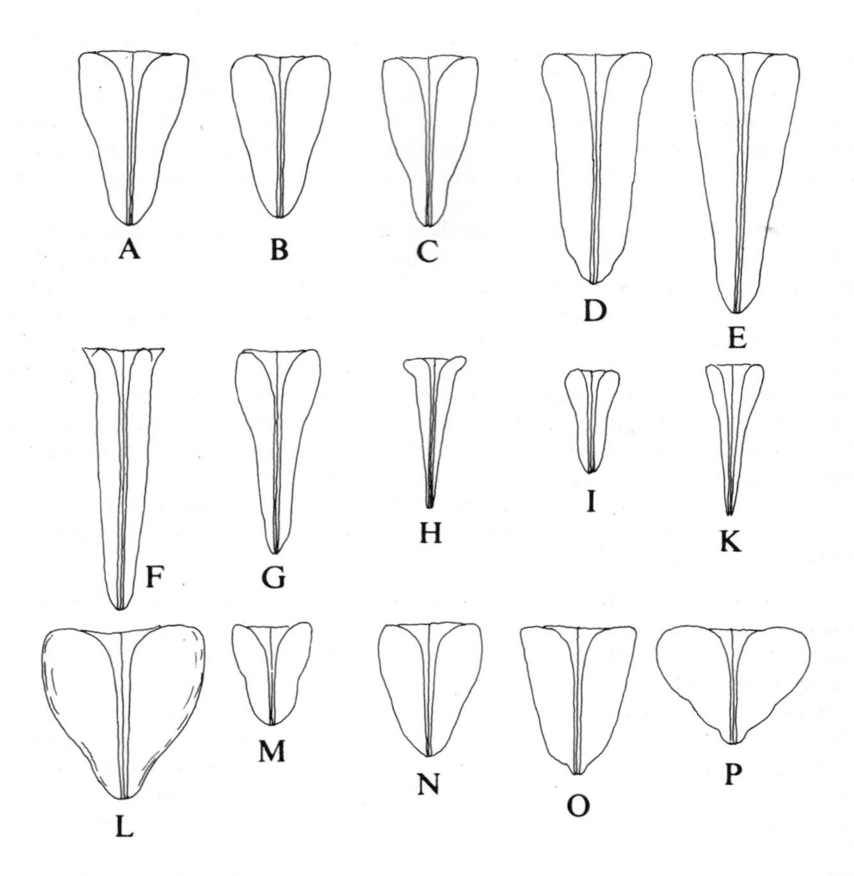

Fig. 45. Shape of apical segment of rostrum in various aphids. – A–E: species on various host plants, but not on Anthemideae or Gramineae; F–K: species on Anthemideae; L–P: species on Gramineae. – A: *Uroleucon (Uromelan) campanulae* (Kalt.); B: *Macrosiphoniella (Asterobium) asteris* (Wlk.); C: *Cryptomyzus ribis* (L.); D: *Rhopalosiphum nymphaeae* (L.); E: *Chaitophorus populeti* (Panz.); F: *Uroleucon achilleae* (Koch); G: *Macrosiphoniella tanacetaria* (Kalt.); H: *Pleotrichophorus glandulosus* (Kalt.); I: *Coloradoa achilleae* H.R.L.; K: *Cryptosiphum artemisiae* Buckton; L: *Sitobion avenae* (F.); M: *Hyalopteroides humilis* (Wlk.); N: *Metopolophium dirhodum* (Wlk.); O: *Rhopalosiphum padi* (L.); P: *Sipha glyceriae* (Kalt.).

occurring on rough surfaces, e.g. on the bark of trunks and branches, have particularly well developed tarsal claws, while aphids found on smooth leaves have relatively short claws, but then a well developed sole bladder at the tip of the tibia (Fig. 25).

Rostrum is very long in species on trunks with rough bark, sometimes even longer than the body, to make it reach the bottom of the crevices, where the stylets shall be inserted, e.g. in *Stomaphis* on oak and in first instar nymphs of *Prociphilus* spp. born on conifers.

The shape of the apical segment of rostrum of aphids associated with certain plant taxa is typical of these species, even if they belong to different systematic groups. The shape is ascribable to convergent evolution, but the advantage of the adaptation is not clear. Aphids on grasses have a relatively short, broad, and obtuse apical segment of rostrum, irrespective of what family they belong to (Fig. 45 L–P), and aphids living on overground parts of plants belonging to the *Anthemis*-group within Compositae (including *Artemisia, Achillea, Tanacetum* a.o.) have a slender, almost stylet-shaped, apical segment of rostrum with slightly concave margins (Fig. 45 F–K).

With regard to colours and colour patterns, sometimes also body shape, mimetic adaptation to particular hosts has developed in many species. Aphids living on branches are usually brown, those on leaves often green or yellow. *Brachycaudus (Thuleaphis) rumexicolens* has the same size, shape, and reddish colour as the flowers and fruits of *Rumex acetosella*, its host plant.

Adaptation sometimes means ability to change the physiological condition of the host plant in such a way that the plant is better suited to meet the demands of the aphid, or rather in the ability to take advantage of such physiological change of the plant. Aphids promoting formation of galls or zoocecidies utilize these for protection against enemies — if the galls are closed — and also to obtain better nutritional conditions.

A similar nutritional advantage is obtained by some aphids through virus infection of the host plant. A virus disease may change the physiological condition of the plant and provide better reproduction, e.g. *Myzus persicae* on sugar beet with virus yellows.

Virus transmission

Many plant viruses are transmitted exclusively or mainly by aphids. After having fed on contaminated plants they infect fresh plants. They belong to several species. Such species are some of the most serious pests to cultivated plants even if they do little harm by their very feeding. The green peach aphid *Myzus persicae* is the vector of more plant viruses than any other insect.

Viruses may be transmitted in two ways. Some viruses may be carried by the apices of the stylets, others are absorbed in the gut, passing through the body to the salivary glands, and transmitted to a fresh plant with the saliva. Viruses transmitted by stylets are called stylet-borne (Fig. 46 A), and those transmitted in the latter way are called circulative viruses (Fig. 46 B).

Many aphids are vectors of stylet-borne viruses. The contagium is readily acquired, but also readily lost. The ability to transmit circulative viruses is only obtained by certain aphid species, and only after feeding for some time. It requires an incubation period, and is retained for a long time. Interaction between plant, virus, and aphid is more advanced than in stylet-borne viruses. Some circulative viruses can multiply inside the aphid, e.g. the leaf roll virus of potato.

Dispersal

In contrast to most other flying insects aphids keep the body almost vertical during flight. The fore wings vibrate approximately in the horizontal plane by the mesothoracic muscles. The hind wings are attached by hooklets to the fore wings and follow their movements.

There are two kinds of flight, an initial direct, vertical upward flight, succeeded by a hovering flight with small vertical movements and short tours in the horizontal plane. The latter kind of flight is accomplished only in calm weather. The aphids often end up in a moving air mass, being seized by the current and carried over longer distances than possible by active flight. Aphids are poor fliers, not able to cover more than 3–5 km per hour. As air plankton they may reach an altitude of 4,000 m and be carried more than 1,200 km away.

Just after the last moult alate aphids cannot fly because the integument and wings are still soft. Hardening takes 2–5 hours, but only after 12–24 hours the aphid is ready for vertical flight, provided that certain external conditions are present. The internal conditions are usually present, but individual alatae may differ in this

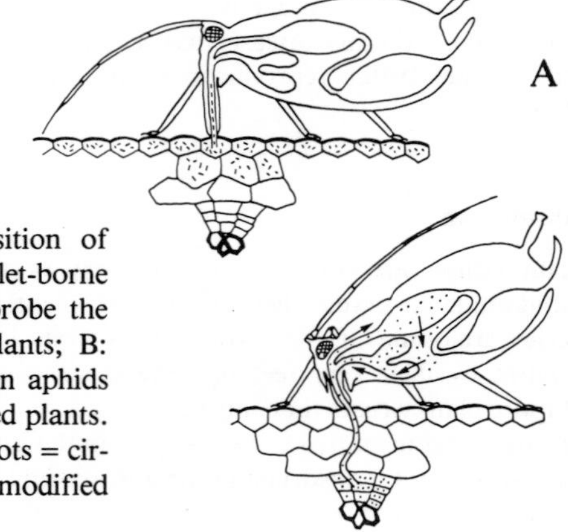

Fig. 46. Mechanisms of acquisition of viruses by aphids. – A: stylet-borne viruses acquired when aphids probe the epidermal tissues of infected plants; B: circulative viruses acquired when aphids feed on phloem tissues of infected plants. – Dashes = stylet-borne virus, dots = circulative virus. (After Dixon, modified after Watson).

respect. Aphids are in the mood to start the flight when feeding and reproduction are inhibited, and this depends on both internal and external factors. Some alatae never fly, while others belonging to the same colony have strong migratory tendencies. Such an individual which may be forced by external factors to stay on the plant where it grew up, does not bear young although it may be full of embryos. Sooner or later it becomes restless and finally attempts to fly, even if the external factors for taking off are not fully favourable.

External factors, e.g. light intensity, temperature, and wind velocity, have been studied only in a few species. Aphids do not start flying in darkness. If the cuticle has not hardened until evening, flight is delayed till next morning. Cloudy, windy, or cold weather may delay the first flight. Occurrence of aphids in strong air currents may be due to the fact that the flight started before the wind increased, or because the flight started from sheltered places.

Moericke (1941, 1955) found that the life of an alata of *Myzus persicae* can be subdivided into four phases, each dominated by a distinct mood. A mood for repose dominates the first phase, comprising the period just after the last moult. A mood for taking off by the wings dominates the second phase; the aphid does not feed, and does not try to feed on the plant where it sits, whether it be a host or not, and it starts flying as soon as the environmental factors are favourable. The flight is regulated by sense stimuli, especially the light from the sky or an artificial light source. In the third phase it tries to alight and probe, but only for a short time. It alights after having flown for about two or three hours, or when the air mass carrying it comes near the ground. Light is necessary for landing, but light of short wavelength from the sky is now less attractive than long-wave reflections from beneath. The aphid is attracted by certain colours, mainly yellow. After alighting it starts flying again, even if it has arrived at a host plant, then settles again for a short while, until the fourth phase, the phase of colonization. In this phase it stays on the plant if this is a host and eventually loses flight ability due to degeneration of the muscles.

Experience indicates that distinct limits between such phases do not always occur. Johnson (1958) showed that a flight period, even only 20–30 seconds, may bring the second phase to an end, and other activities or narcotization with CO_2 can do so. Between hardening of the integument (or a little later) and loss of flight ability, corresponding to Moericke's second and third phases, the aphids seem to be in the mood for taking off when they have rested for some time, and in the mood for landing when they have been active or otherwise have a surplus of CO_2. Until the beginning of the fourth phase they are in the mood to take off equal to the mood finishing the first phase, and their visits to plants do not increase in length. The duration of a visit depends partly on the length of the previous flight period, partly on the environmental factors when they start another flight, and non-hosts are left sooner than hosts. The mood to take off, revived by alighting on a non-host, can only be abolished by an extra flight for at least 20 seconds, not by immediate transfer to a host.

After each landing the aphid will probe the surface with the stylets to taste whether the plant is a host or not. It will probe for a longer time on hosts than on non-hosts and sometimes deposit young. For probing only the epidermis needs be penetrated. This takes seconds, while reaching the sieve tubes (the phloem) may take more than half an hour (Hille Ris Lambers 1972).

If alate aphids do not get an opportunity to fly, they calm down after some days, and bear young. If after a short flight aphids are placed in darkness for 24 hours the same effect as a flight of several hours is obtained; they will feed and reproduce if placed on a host, but not on a non-host.

In Johnson's experiments mostly short flights, artificially brought to an end, were studied. Moericke's phases can be observed under natural conditions, since the first flight is usually very long, the aphids being seized by the wind and passively carried along for a long time.

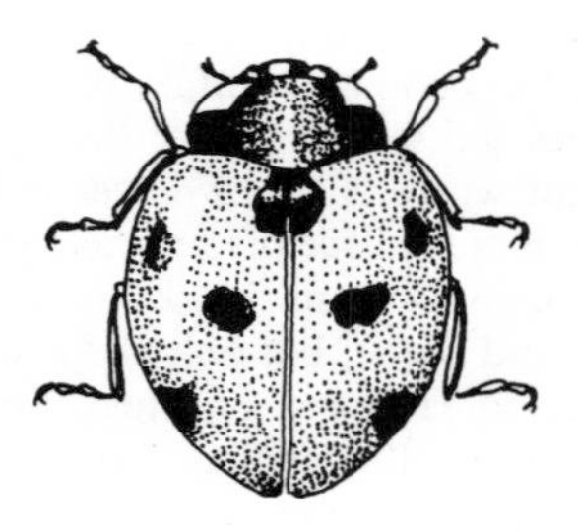

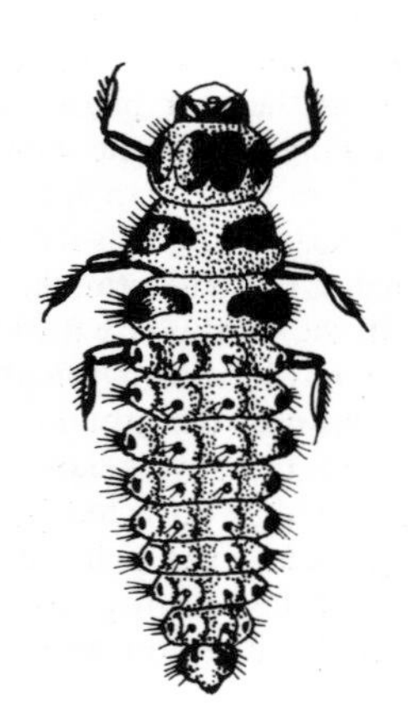

Fig. 47. *Coccinella septempunctata* L., imago and larva. (After Bovien & Thomsen, redrawn).

Predators and parasites

The great multiplication of aphids due to rapid development and parthenogenesis is counterbalanced in nature by a high rate of mortality, which is partly due to failure of the alatae to find host plants, partly to attacks by predators and parasites. This balance does not keep populations on a constant level. They grow rapidly to a high level, before departure of alatae, predation and parasitism drastically reduce them.

The most important predators in Scandinavia are ladybirds (Coccinellidae) and their larvae, especially *Coccinella septempunctata* (Fig. 47), lacewings (Neuroptera) and their larvae, e.g. *Chrysopa perla*, larvae of hover-flies (Syrphidae), larvae of cecid flies (Cecidomyiidae), and predatory bugs as Anthocoridae a.o. (Heteroptera).

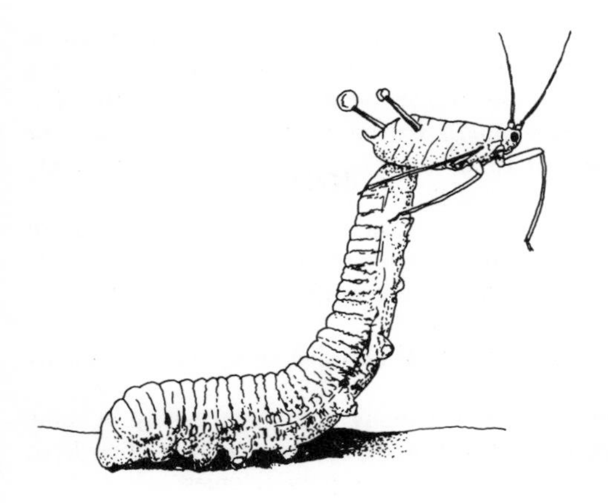

Fig. 48. Syrphid larva attacking an aphid which produces droplets (with alarm pheromone) from the siphunculi.

Adult ladybirds can eat 20 aphids or more a day. It has been calculated that 200–500 aphids are eaten during the larval life, and 1000–2500 in the adult life. A neuropterous larva can eat more than 100 aphids, a syrphid larva between 200 and 1000. Syrphid larvae are more common in aphid colonies than any other predators.

Less important enemies are some more or less omnivorous animals as spiders, earwigs, rove beetles, lizards, and birds (tits, sparrows, swallows a.o.). Some digger wasps store paralyzed aphids in subterranean nests as food for their larvae.

The most important parasites are parasitic wasps (Aphidiidae a.o.) and fungi (*Entomophthora* a.o.). Larvae of parasitic wasps develop within single individuals. The female wasp oviposits into the body of an aphid, usually an immature specimen. The larva feeds on the body contents, leaving the vital organs until last, so that the aphid remains alive until shortly before the larva is ready to pupate. The aphid dies before being full-grown if parasitized early in life, but may become an adult if

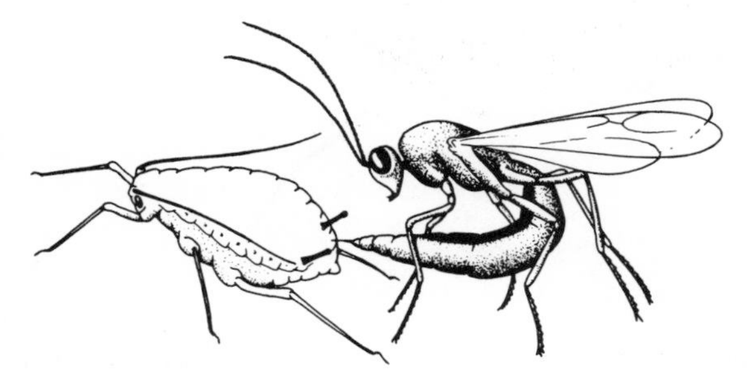

Fig. 49. Parasitic wasp *(Aphidius)* ovipositing in an aphid. (After Webster).

parasitized at a later stage. It looks distended and often shining just prior to its death, and often the larva can be observed through the hyaline skin. The larva spins a cocoon of silk and pupates, usually inside the dead aphid (Fig. 50). The larva of *Praon* forces its way out on the aphid's ventral surface, and constructs the cocoon between the aphid and the surface of the plant so that the dead empty aphid becomes fastened to the leaf by a conical socket (Fig. 51).

The adult wasp leaves the cocoon through a circular hole at the posterior end of the mummified aphid, or — in the case of *Praon* — in the socket.

Hyperparasitic wasps oviposit in larvae of parasitic wasps in aphids. Such wasps are not aphid enemies, but enemies of their enemies.

Red mites (Thrombidiidae) may occur as external parasites of aphids, apparently without being very harmful.

Parasitic fungi are more dangerous. They kill many aphids, especially in rainy summers. An attacked aphid becomes brownish and — after death — powdered with spores. The epidemic disease spreads quickly throughout the colonies.

Aphids seem to be an easy source of food. They are not totally defenceless, but their defence is passive. An enemy may alarm a whole colony, and some aphids may succeed in withdrawing their stylets from the plant tissues and escape. Aphids may warn other aphids by sudden movements of body, antennae, or hind legs, or by secreting siphuncular droplets containing an alarm pheromone. Some aphids escape by letting themselves drop to the ground. Still more passive, but probably more effective defensive measures are formation of closed galls or presence of poisonous substances in the body, which reduce the fecundity of ladybirds, which eat them (Blackman 1967). Such substances are found e.g. in *Megoura viciae* and *Aphis sambuci.* Mimetic colours may help aphids prevent attention of some enemies.

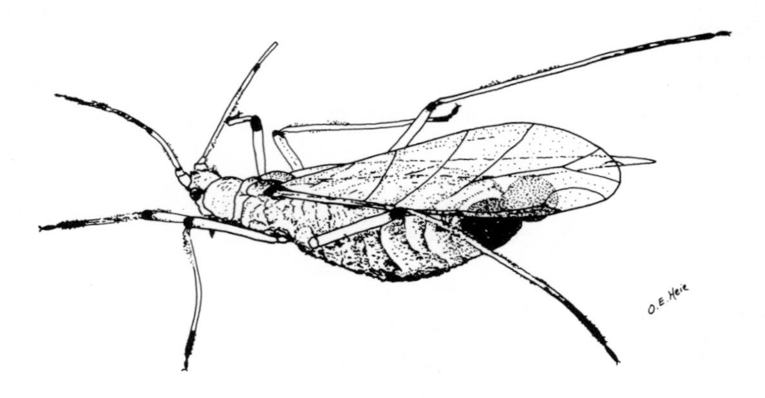

Fig. 50. *Euceraphis punctipennis* (Zett.) killed by *Trioxys compressicornis* Ruthe.

Relation to ants

Many aphid species are myrmecophilous. They are visited and attended by ants which use their excrements, the honeydew, as food source. Honeydew is attractive to ants because aphids take up more sugar than they need, and have to give off the surplus.

It is emphasized that not all aphid species are visited by ants, and there are several degrees of interdependence. Strongly wax-producing species, e.g. *Eriosoma lanigerum*, are avoided by ants, because the excrements are not attractive. Mordvilko was the first to realize that the composition of the excrements to some extent is dependent on the quantity of wax secretion.

Aphids living scattered all over the host plant are not either myrmecophilous. It is characteristic of species depending on ants that they form dense colonies, even if closely related species, which are not visited by ants, do not form colonies. The *Pterocallis* spp. living on *Alnus glutinosa* is a good example, *P. alni* living scattered without ants, *P. maculata* living in colonies visited by ants. It should however, be added that many aphids, which are independent on ants, can also live in colonies.

Aphids, which throw their excrements far away atomized into tiny droplets, are not visited by ants, but the ants, together with bees, flies, and other insects, sometimes collect the honeydew from the leaves underneath, when great attacks by colonies give rise to a drizzle of tiny excrement drops. Many of these aphids have a long cauda and long siphunculi.

Some of the myrmecophilous aphids are not totally dependent on ants, but seem to thrive better if they are visited by ants. They do not react to the visitors in the

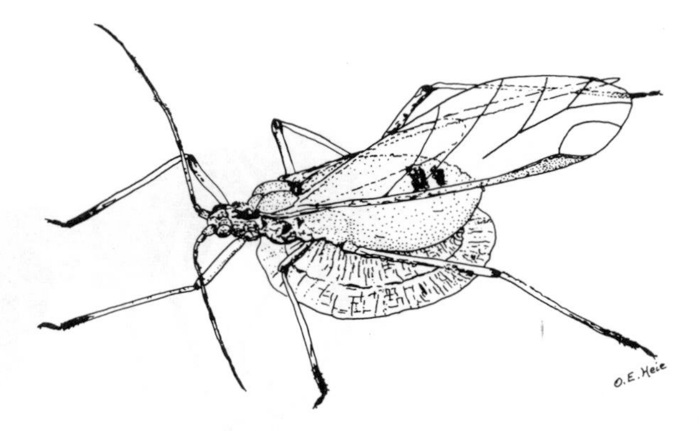

Fig. 51. *Euceraphis punctipennis* (Zett.) killed by *Praon flavinode* (Hal.). The cocoon of the wasp is seen under the emptied body.

same way as to enemies. When touched by the antennae of the ants they do not make alarm movements or siphuncular droplets, and they deliver their excrements as drops instead of throwing them away so that the ants can easily take up the honeydew. The ants benefit from the aphids by getting food. Honeydew is an important food to many ants, both Formicinae (*Lasius, Formica* a.o.) and Myrmecinae (*Myrmica, Tetramorium* a.o.). Aphids also benefit from ants.

Aphids get rid of their sticky excrements and may be tranquilized so that they can save energy for feeding and reproduction. The loss connected with production of alatae is also smaller because ants promote production of apterae. Ants protect the colonies against many enemies, but parasitic wasps and some predators nevertheless are often able to attack aphids attended by ants.

The association with ants is still more advanced in some subterranean aphids, e.g. many species within Pemphigidae. These species or some of their generations are totally dependent on ants. Some have hairs around the dorsally placed anus (Fig. 223), and the honeydew droplet is kept there until removed by an ant. The ants protect the aphids by building earth galleries around the colonies, and in some cases they carry aphids to fresh host plants or keep them alive in their nest during winter and take them out again in spring. This behaviour has been recorded for *Lasius flavus,* a specialist on animal husbandry among ants.

The association is a kind of symbiosis or mutualism. Among the myrmecophilous insects these aphids belong to the trophobionts.

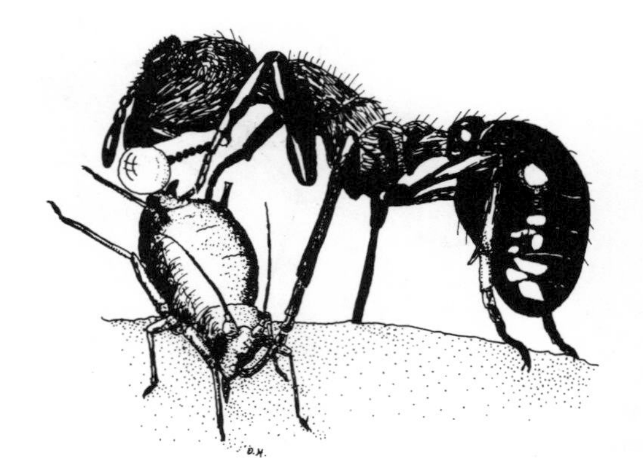

Fig. 52. *Formica* taking up honeydew. (Original drawing on the basis of a photo taken by Elvig Hansen).

Zoogeography

At present 579 species of Aphidoidea are known from Denmark and Fennoscandia, but the number of species may be much higher. Many records have been added rather recently. In Great Britain, one of the more intensively explored areas, 535 species have been recorded, although the area is only about one-fifth of the Scandinavian countries. Szelegiewicz (1978) lists 679 species from Poland.

Most species are western palaearctic or holarctic and widespread in Europe. The similarity to the aphid fauna of Great Britain is considerable, since 93% of the Scandinavian species have been found there. The remaining 7% are partly arctic (e.g. *Betulaphis pelei*) or eastern (e.g. *Tinocallis saltans)* species, partly rare (e.g. *Coloradoa campestrella*) or little known, recently described species, including endemic ones. Some of the latter, e.g. *Eriosoma (Schizoneura) anncharlotteae*, are probably more widely distributed.

The number of species known in Fennoscandia and Denmark represents about 15% of the world fauna. This is a great part compared with the area and is at least partly due to the fact that several kinds of landscapes and vegetation occur: lowland and mountains, agricultural land, dunes, moors, deciduous forests, coniferous forests, and tundra, areas with temperate climate, including both oceanic and continental ones, and also areas with arctic climate.

The number of species hitherto recorded declines to the north; 83% occur in southern Scandinavia (Denmark, and the districts Sk., Bl., and Hall. in Sweden), while only 14% occur within the arctic circle (the districts T.Lpm., Nn, TR, F, Le, and Li).

Aphidoidea is richer in species in temperate climates than in subtropical and tropical regions. The general life cycle with several generations of parthenogenetic viviparous females followed by formation of sexuales, which produce diapause eggs, must be regarded as an adaptation to climates with cold winters, or at least to climates with seasonal changes of temperature and daylength. Arctic faunas are poorer, partly due to reduced selection of suitable food sources, partly directly to climatic conditions.

The occurrence of a species generally corresponds to the distribution of the plants on which its life cycle depends, but in some cases aphids range farther north than the primary hosts. The occurrence of e.g. *Myzus persicae* and some Pemphigidae depends on the distribution of the plants on which anholocyclic wintering is possible year by year.

The northern distribution limit of many species passes through Scandinavia, because they live on plants which have their northern distribution limit here. The distribution of *Phyllaphis fagi* is a typical example. The presence of this species obviously depends on the presence of its host (beech – *Fagus silvatica*), and not on certain climatic conditions, since it has been found also outside the natural range of beech in places where beech has been planted, e.g. in the Helsinki Botanical Garden (Fig. 53). The northern border of some other species may depend on climatic factors, if the host plants extend farther north than the aphids. *Macrosiphum funestum* is common in southern Denmark, but rare in northern Denmark, and is not known from Fennoscandia although its hosts, *Rubus* spp. of the *fruticosus-caesius* group, extend north to southern Norway and central Sweden (Fig. 54).

An analysis of the distribution of the about 600 species and subspecies is not given here. Brief comments on the known distribution of species and subspecies are given after each description. Spreading by man or by air currents has extended the distribution, and the origin of many species can only be ascertained by the geographic origin of the plant which seems to be the original host.

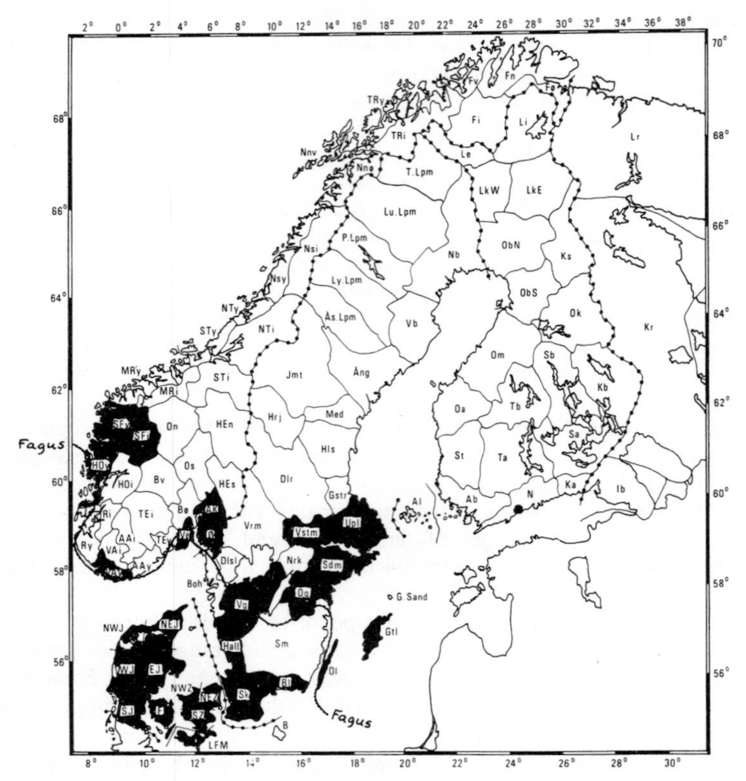

Fig. 53. Distribution of *Phyllaphis fagi* (L.) in Scandinavia. – Black = districts from which the species is recorded; Finland: one record from Helsinki. The natural northern limit of beech (*Fagus silvatica*) is based on Lid (1963). Beech occurs in gardens also north of this limit.

Several species, mostly cosmopolitan, are not indigenous but introduced together with cultivated plants. Some are found only in gardens and plantations (*Eriosoma lanigerum, Acyrthosiphon caraganae* a.o.), a few exclusively on indoor plants (*Cerataphis orchidearum, Dysaphis tulipae, Idiopterus nephrelepidis, Myzus dianthicola* a.o.). The geographic origin of some species possibly introduced more than a century ago is obscure.

Some species from other zoogeographic regions were added recently: *Illinoia (Masonaphis) lambersi* (from North America), *Rhopalosiphum rufulum* (maybe also from North America, maybe from East Asia, where its secondary host, *Acorus calamus,* originates), *Myzus ascalonicus* (origin unknown) a.o. They recently arrived at other parts of Europe and are now abundant locally.

A few species known only from traps may not be true members of the Scandinavian fauna, but may have been carried by the wind from far away places, e.g. *Trichosiphonaphis (Xenomyzus) corticis*, from mountains in Central and SE Europe.

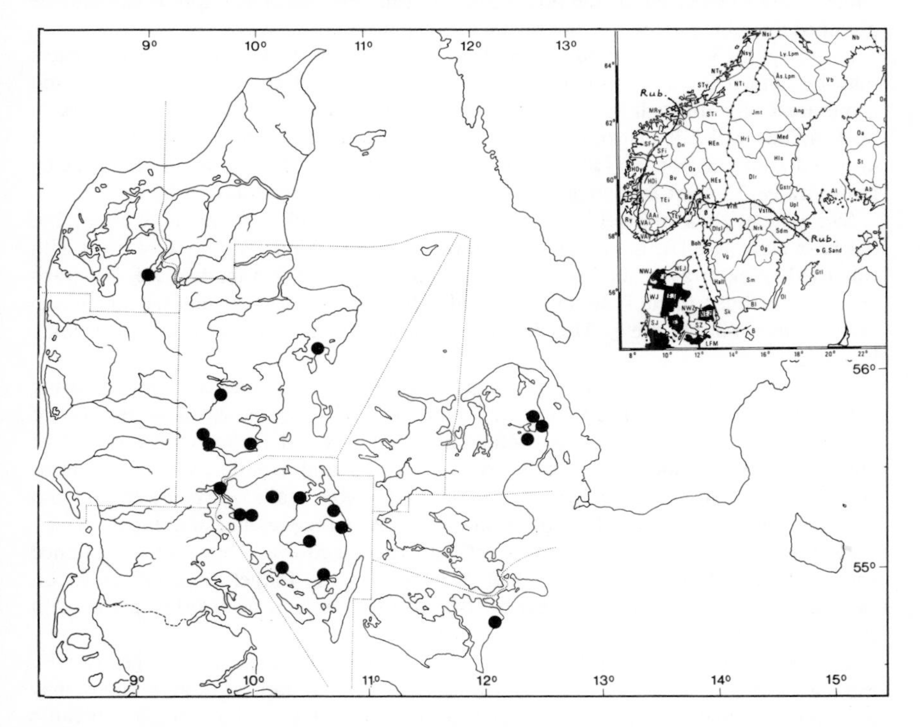

Fig. 54. Distribution of *Macrosiphum funestum* (Macch.) in Denmark and N Germany. The northern limit of *Rubus* spp. of the *fruticosus-caesius*-group is given on the inset map, based on Lid (1963).

Collecting

The equipment comprises glass tubes or plastic containers with lids (e.g. cork or cotton), forceps, pocket lense, a board of uniform colour with an incision so that it can be placed close to the base of a plant stem or tuft of grass, a scoop or small shovel, and perhaps also a pair of scissors (to cut off shoots with aphid colonies), a fine brush, and a butterfly net with a strong frame. The containers shall be sufficiently large to contain whole leaves or parts of stems or roots.

Aphids should be looked for on all parts of plants, especially on terminal shoots and undersides of leaves, but also on roots, rhizomes, trunks, branches, uppersides of leaves, and among flowers and fruits.

Subterranean aphids can be found by digging up plants or part of roots (if trees), which should be placed on the board and carefully cleaned. Such aphids can also be obtained by pulling up plants, but then several aphids may be lost.

Some aphids living on plant parts above ground are easily detected, while others may be hidden in flowers or leaf sheaths. Most species feeding on Gramineae and Cyperaceae are difficult to observe in situ, even if they occur in large numbers. Such aphids will fall off and are then easy to detect, if the plants are shaken and knocked over the board. This should be framed to prevent the aphids from being blown away. They are easy to pick up with a slightly moistened brush or forceps.

Sweeping with a butterfly net quickly reveals the aphids in the vegetation, but the host plant is difficult to identify. Traps of various kinds, e.g. yellow trays with water (Moericke trays), suction traps, and light traps, are also inadequate, but traps are recommended to find the amount of aphids in air masses, calculating population fluctuations, migration etc. Also for control measures and on expeditions traps may be useful.

Many species may be discovered on plants growing in ants' nests or by following ant roads from nest to food sources. Aphids on branches or leaves at the top of tall trees are difficult to collect. Some can be discovered through observation of ants on the trunks and then be collected with a stick and a butterfly net. Ladybirds and other predators, curled leaves, discoloured leaves, galls, wax secretions, and honeydew on the ground and on lower leaves can reveal the presence of aphids.

Aphids on moss are difficult to detect with the naked eye. They can be obtained with a Berlese funnel. The moss shall be placed in the funnel on a net under an electric bulb. This method can also be used to expel aphids from other plants with closely sitting small leaves.

Aphids collected on their host plants shall be brought home alive together with the leaves, stems, or roots, on which they have been found, so that nymphs may get time to develop into adults. The live aphids shall be closely studied because colours will disappear after preservation. Predatory insects shall be removed as soon as possible. Parasitized aphids need not be removed. The parasites can be reared and stored in alcohol for identification later.

Aphids can be kept alive on pieces of stems or whole leaves for several days — even for weeks if fresh stems or leaves are supplied now and then — in tubes kept suitably moist. The host range can be determined if the plant material is replaced by material of other plant species, but it is more adequate to rear several generations of the aphids on potted plants. To rear aphids dependent on ants is difficult if not combined with a formicarium.

The aphids are killed in glass tubes with 80–96% alcohol. They are transferred from the container or directly from the plant to the tube with a slightly moistened brush or forceps. The glass tube shall be closed by a piece of cotton wool to avoid air bubbles in the tube. Aphids become brittle in alcohol and may lose antennae and legs if kept in a glass tube with air bubbles which is shaken. Containers and glass tubes shall be labelled. The labels can show numbers referring to a journal indicating: locality, date, host plant (or number referring to the herbarium), and perhaps also the colour of the aphids, their localization on the plant, presence of ants and enemies, and other observations made in the field or in the laboratory until the aphids were killed. The glass tubes should be kept in alcohol in closed jars until the preparation will take place.

Preparation

The following method is that presented by Hille Ris Lambers (1951).

The aphids are heated in 96% alcohol (with a piece of pumice or the head of a burnt match to avoid boiling over) in a water bath just beyond the boiling point for 5–10 minutes, unless they have been kept in alcohol for one month or more.

The alcohol is poured out of the glass tube and replaced by 10% caustic potash (KOH). Then the tube with the aphids in KOH is heated for 1–5 minutes in water bath so that the KOH is just about to boil. The bigger and the darker the aphids are, the longer heating time is needed.

After cooling KOH is decanted. Alcohol is added to the glass tube and poured out again several times to remove any trace of KOH.

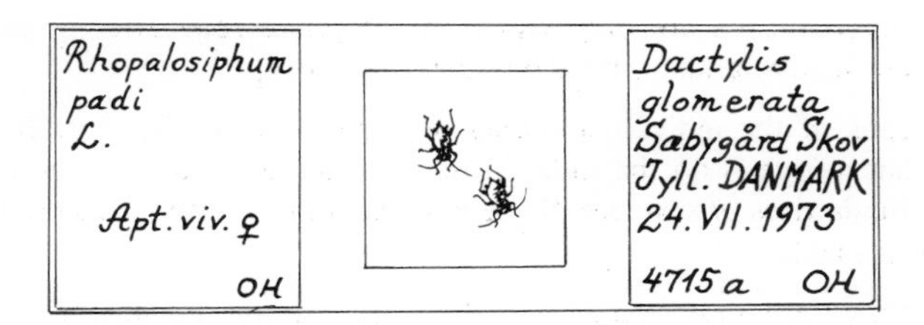

Fig. 55. Slide with mounted aphids and correct labelling.

Then some drops of chloralphenol, a saturated solution of chloralhydrate in phenol (phenolum liquefactum), are added to the glass tube containing the aphids. This is a strong poison and must not be poured into the plug hole after use. Inhalation of the vapours must be avoided. The glass tube is heated again in water bath at 100°C for 5–10 minutes in a fume cupboard with automatic exhaustion.

The aphids are now cleared. They can be kept for a long time in the cold chloralphenol if stored in darkness. After cooling they are transferred to a glass slide with a drop of a mounting medium consisting of 12 g clean gum arabic + 6.5 ml pure glycerol + 20 g chloralhydrate + 20 ml distilled water, arranged in the desired position (by means of needles under the microscope) and covered with a coverslip. A single slide can contain more than one specimen, depending on the size of the aphids.

The slide must be placed in a horizontal position for a month (if not dried in a heating cupboard) and then sealed, for instance with clear nail varnish to avoid exsiccation and air between slide and coverslip.

The labels shall indicate locality, date, host plant, and name or initials of the collector, also journal number. The name of the aphid and the name of the person who identified it shall be added sooner or later, preferably on a label reserved for this purpose.

Labels made of cardboard instead of paper make storage of the slides easy, as they can be stacked like index cards in boxes measuring 28×92×130–140 mm, and the risk of ruining them is also smaller. The boxes can be arranged like books on a shelf, with the slides in a horizontal position with upside down to avoid dust.

Identification

Some work has to be done before the keys can be used. Identification is difficult because:

1) Diagnostic characters are only visible in the microscope after preparation. The aphids must be hyaline and mounted on slides (see above).

2) The morphs of the individual species are usually very different. It is necessary to know what morph is on the slide. Most keys cannot be used if the material only consists of fundatrices or sexuales. Some keys can only be used if apterous viviparous females are available.

3) Some species are morphologically alike and cannot be identified if the host plant is unknown. In some cases live aphids have to be transferred to certain other plants to show if these are also accepted as hosts.

4) The range of intraspecific variation within the same morph may be greater than that given in keys and descriptions so that rare extreme variants cannot be identified. It is recommended to study as many specimens as possible and get an idea of the range of variation within the whole sample.

5) Diagnostic characters are mainly based on exact measurements of the antennal segments, hairs, and other morphological features. Identification therefore is facilitated if measurements of general importance are taken and listed in a table for each of the specimens.

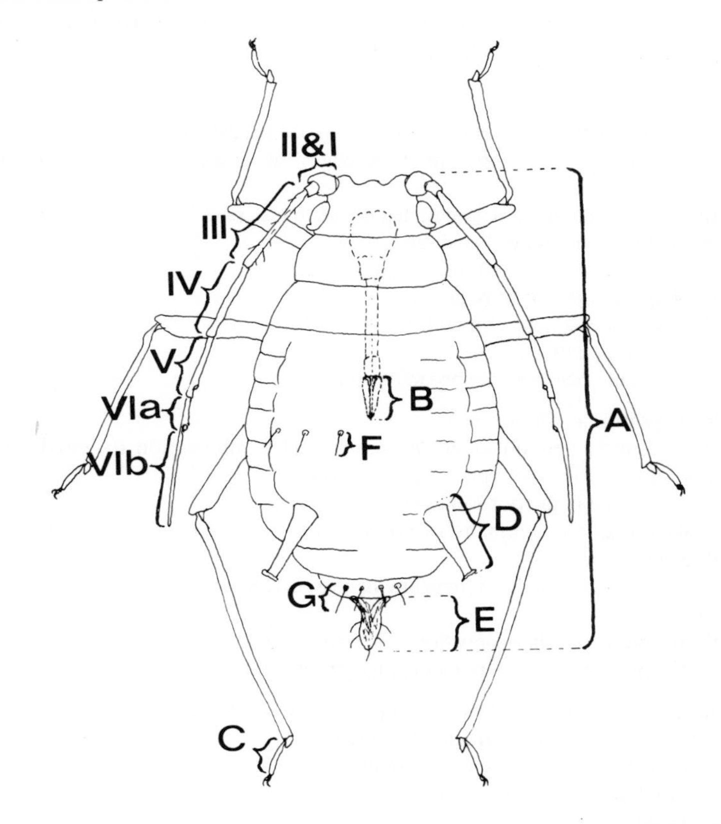

Fig. 56. Measurements. – A = body length; B = length of apical segment of rostrum; C = second segment of hind tarsus (2sht.); D = siphunculus length; E = length of cauda; F = length of hair on abd. tergite III; G = length of hair on abd. tergite VIII; I–VI = length of antenna; VIa = basal part of ant. segm. VI; VIb = processus terminalis.

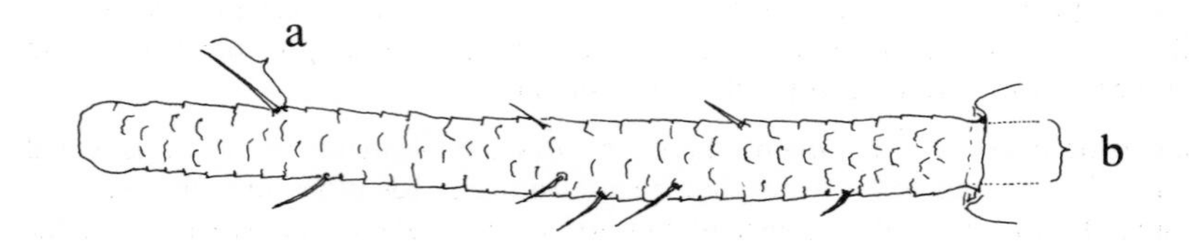

Fig. 57. Antennal segment III of apterous viviparous female of *Aphis fabae* (Scop.). – a = length of longest hair; b = diameter of segment at basal constriction (= III bd.).

The most important measurements (in mm to two decimals) are:

Length of body from frons (or fore edges of frontal tubercles) to apex of cauda.

Length of each antennal segment; length of basal part of ultimate segment and processus terminalis should be listed separately.

Length of antenna (= the sum of the lengths above).

Length of apical segment of rostrum.

Length of second segment of hind tarsus (= 2 sht.).

Length of siphunculus.

Length of cauda (including the broad basal part).

The following ratios are important:

Length of longest hair on antennal segment III in proportion to length of basal diameter (width) of antennal segment III (or III bd.).

Length of antenna in proportion to length of body.

Length of processus terminalis in proportion to length of basal part of ultimate antennal segment.

Length of apical segment of rostrum in proportion to length of second segment of hind tarsus.

Length of siphunculus in proportion to length of body.

Length of siphunculus in proportion to length of cauda.

It is important to count the following morphological components or details:

Secondary rhinaria on each antennal segment.

Accessory hairs on apical segment of rostrum.

Hairs on first tarsal segment of each leg.

Hairs on cauda.

Key to families of Aphidoidea

(partly adapted from Shaposhnikov (1964) and Stroyan (1977))

Apterous viviparous females

1 Head capsule and pronotum fused; compound eyes situated near middle of the apparent "head" (Fig. 4 B) 2

– Head capsule and pronotum separated, at least laterally behind the eyes 6

2 (1) Siphunculi well developed, sometimes very long, often haired (Fig. 58). Eyes large and multifacetted or small and three-facetted. .... **Greenideidae** (not in Scandinavia)

– Siphunculi small, pore-shaped, not haired. Eyes small, with three facets (Fig. 15 B) 3

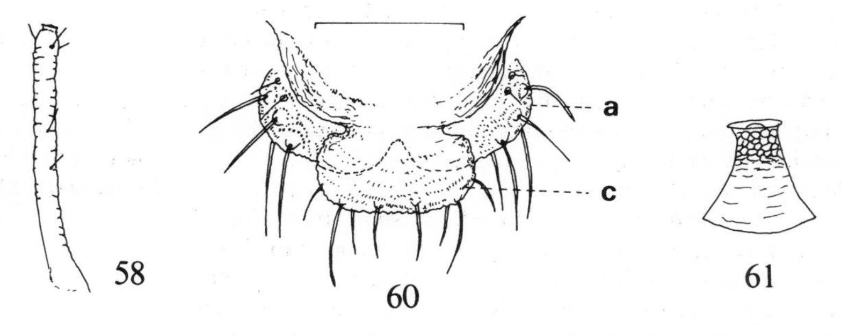

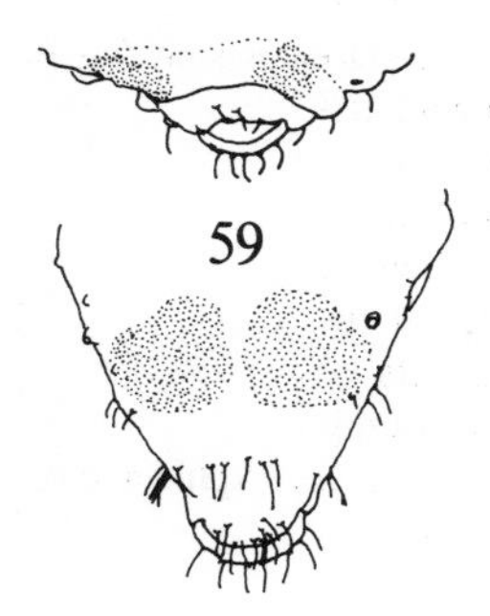

Fig. 58. Siphunculus of *Cervaphis* (Greenideidae). (After Hille Ris Lambers).

Fig. 59. Abdominal wax glands of *Phloeomyzus*, aptera above and alata below. (After Shaposhnikov, redrawn).

Fig. 60. The bilobed anal plate (a) of the alate viviparous female of *Hamamelistes betulinus* (Horv.), partly covered by the knobbed cauda (c). (Scale 0.1 mm).

Fig. 61. Siphunculus of *Periphyllus* (Drepanosiphidae: Chaitophorinae).

3 (2) Cauda subtriangular (Fig. 84) .... **Mindaridae** (p. 83)
– Cauda knobbed (Figs. 60, 100) or broadly rounded (Fig. 101) 4
4 (3) Wax glands present only on abdominal tergite VII, arranged in two large groups (Fig. 59) .... **Phloeomyzidae** (not in Scandinavia)
– Wax glands absent, or if present, then not only on abdominal tergite VII 5
5 (4) Anal plate bilobed (Fig. 60). Antennae and legs poorly developed, concealed beneath body. .... **Hormaphididae** (p. 87)
– Anal plate not bilobed. Antennae and legs well developed, not concealed beneath body. .... **Thelaxidae** (p. 93)
6 (1) Processus terminalis shorter than 0.5 × basal part of ultimate antennal segment. Cauda broadly rounded (Fig. 31 C) 7
– Processus terminalis longer than 0.5 × basal part of ultimate antennal segment; if shorter then cauda knobbed (Fig. 31 E) 9
7 (6) Apical segment of rostrum not distinctly subdivided into two parts (Figs. 165, 212). Eyes three-facetted (Fig. 15 B), or, if multifacetted, then siphuncular pores absent (except in some alatiform apterae on *Ulmus*). Marginal tubercles absent. Hind tarsus much shorter than 0.5 × length of hind tibia. .... **Pemphigidae** (p. 111)
– Apical segment of rostrum distinctly subdivided into two parts (Figs. 72, 73, 116). Eyes multifacetted, *or,* if three-facetted, then *either* siphunculi and marginal tubercles present *or* hind tarsus about 0.5 × length of hind tibia or longer. 8
8 (7) Marginal tubercles present on prothorax and some of the abdominal segments (Fig. 117) .... **Anoeciidae** (p. 100)
– Marginal tubercles absent. .... **Lachnidae** (p. 81)
9 (6) Cauda knobbed, broadly rounded, or short triangular; if broadly rounded, then *either* siphunculi stump-shaped and with reticulate sculpture (Fig. 61), *or* antenna 5-segmented and marginal tubercles absent, *or* secondary rhinaria narrow and transverse; if cauda short triangular, then anal plate with a deep cleft or emargination posteriorly. .... **Drepanosiphidae** (p. 77)
– Cauda finger-shaped, ensiform, tongue-shaped, elongate triangular, short triangular, helmet-shaped, or broadly rounded; if helmet-shaped or broadly rounded, then siphunculi without reticulate sculpture and not stump-shaped, and antenna 6- or 4-segmented (if 5-segmented, then large marginal tubercles present on pronotum and some other segments of body). Secondary rhinaria, when present, more or less round. Anal plate not emarginate or cleft posteriorly .... **Aphididae** (p. 79)

Alate viviparous females

1 Fore wing with radial sector originating from base of pterostigma (Fig. 80) .... **Mindaridae** (p. 83)
– Fore wing with radial sector originating from middle or distal part of pterostigma, or radial sector absent. 2
2 (1) Siphunculi well developed, haired (but not pore-shaped and placed on hairy cones) (Fig. 58). Cauda broadly rounded or conical, sometimes with a slender, projecting apical part (the stylus). .... **Greenideidae** (not in Scandinavia)
– Siphunculi absent, pore-shaped (and then sometimes placed on low hairy cones (Fig. 30 K)), or well developed, with or without hairs; if well developed and with hairs, then cauda not broadly rounded or conical. 3
3 (2) Processus terminalis shorter than 0.5 × basal part of ultimate antennal segment. Cauda broadly rounded (Fig. 101) or knobbed (Figs. 60 and 100); if knobbed, then antenna with less than 6 segments. 4
– Processus terminalis longer than 0.5 × basal part of ultimate antennal segment, or, if shorter than 0.5 × this length, then cauda knobbed and antenna 6-segmented. 9
4 (3) Wings roof-like in repose; mesothoracic lobes well developed (Fig. 21 A), or, if wings flat in repose and mesothoracic lobes

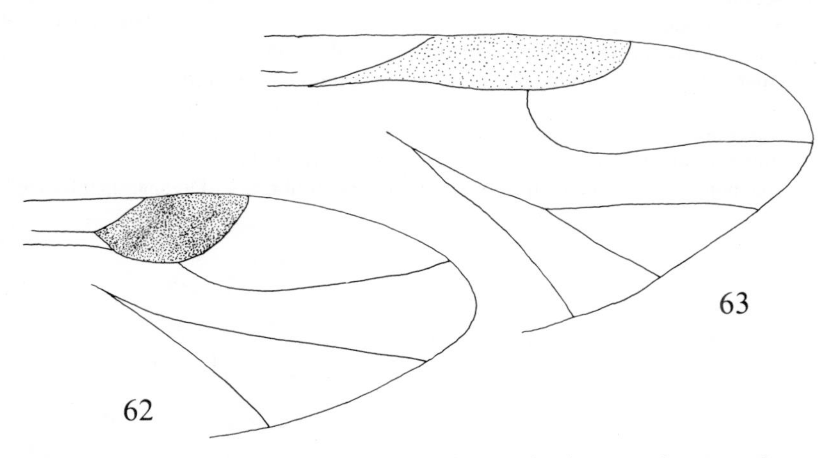

Figs. 62, 63. Apex of fore wing, pigmentation omitted except in pterostigma. – 62: *Anoecia major* Börner; 63: *Lachnus iliciphilus* (del Guerc.).

little developed, then with wax gland plates present marginally and spinally on all abdominal tergites. 5

– Wings flat in repose; mesothoracic lobes little developed (Fig. 21 B). If wax gland plates present, then only on abdominal segment VII or in the form of a border round the body 7

5 (4) Apical segment of rostrum not distinctly subdivided into two parts (Figs. 165, 212). Siphunculi pore-shaped or absent. Hind tarsus much shorter than 0.5 × length of hind tibia. Media of fore wing with one fork or unbranched, only exceptionally with two forks in individual specimens. .... **Pemphigidae** (p. 111)

– Apical segment of rostrum distinctly subdivided into two parts (Figs. 72, 73, 116). Siphunculi conical, pore-shaped, or absent; if absent, then hind tarsus about as long as 0.5 × length of hind tibia or longer. Media of fore wing with one or two forks, rarely unbranched. 6

6 (5) Pterostigma less than 4 times as long as wide, usually black (Fig. 62). Marginal tubercles present. .... **Anoeciidae** (p. 100)

– Pterostigma 4–20 times as long as wide, usually brownish (Fig. 63). Marginal tubercles absent. .... **Lachnidae** (p. 81)

7 (4) Antenna 6-segmented, without secondary rhinaria. **Phloeomyzidae** (not in Scandinavia)

– Antenna 5-segmented, with secondary rhinaria. 8

8 (7) Secondary rhinaria ring-like (Fig. 97). Anal plate bilobed (Fig. 60). .... **Hormaphididae** (p. 87)

– Secondary rhinaria subcircular (Figs. 103 and 110). Anal plate rounded. .... **Thelaxidae** (p. 93)

9 (3) Cauda knobbed (Fig. 31 E), broadly rounded, or short triangular; if broadly rounded, then *either* siphunculi stump-shaped and with reticulate sculpture (Fig. 61), *or* antenna 5-segmented and marginal tubercles absent, *or* secondary rhinaria narrow and transverse; if cauda short triangular, then anal plate with a deep cleft or emargination posteriorly. ... **Drepanosiphidae** (p. 77)

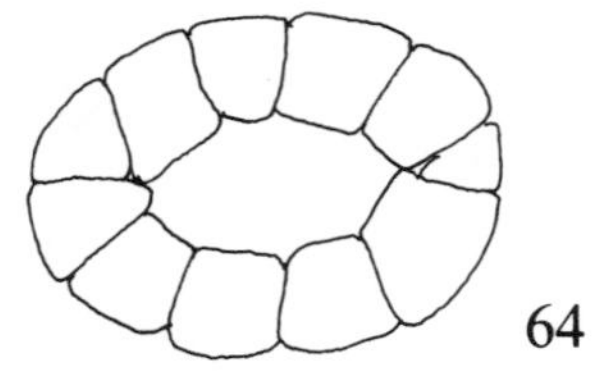

65

Figs. 64, 65. Abdominal wax gland plates – 64: *Eriosoma* (Eriosomatinae); 65: *Gootiella* (Pemphiginae).

– Cauda finger-shaped, ensiform, tongue-shaped, elongate triangular, short triangular, helmet-shaped, or broadly rounded; if helmet-shaped or broadly rounded then siphunculi without reticulate sculpture and not stump-shaped, and antenna 6- or 4-segmented (if 5 segmented, then large marginal tubercles present on pronotum and some other body segments). Secondary rhinaria, when present, more or less round. Anal plate not emarginate or cleft posteriorly. .... **Aphididae** (p. 79)

Keys to lower taxa of Mindaridae, Hormaphididae, Thelaxidae, Anoeciidae, and Pemphigidae are given later in this volume. Keys to subfamilies and tribes of families to be treated in later volumes are given below.

Key to subfamilies and tribes of Drepanosiphidae

Apterous and alate viviparous females

1 With two rudimentary gonapophyses (Fig. 66 B). Basal part of second segment of rostrum usually with sclerotized wishbone-shaped arch (Figs. 67, 68). Siphunculi never reticulate. ... **Phyllaphidinae** – 2

– With three or four rudimentary gonapophyses. Basal part of second segment of rostrum without sclerotized arch (Fig. 69). Siphunculi sometimes reticulate (Fig. 61) 3

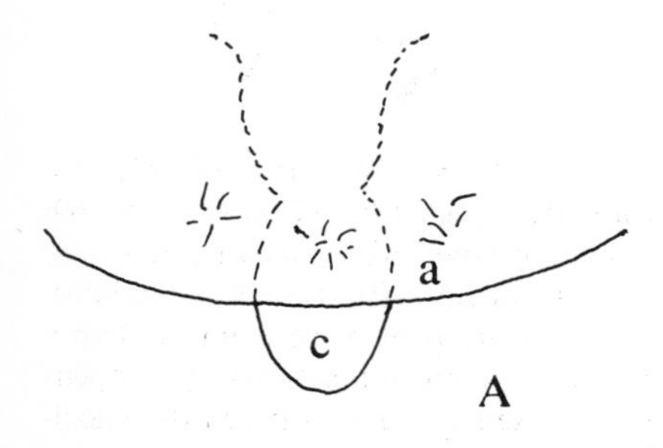

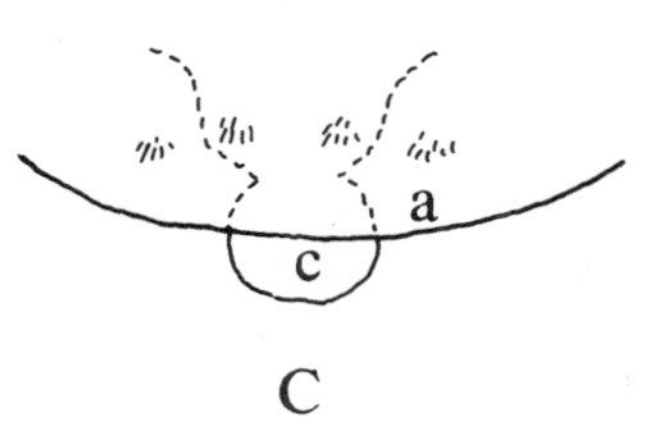

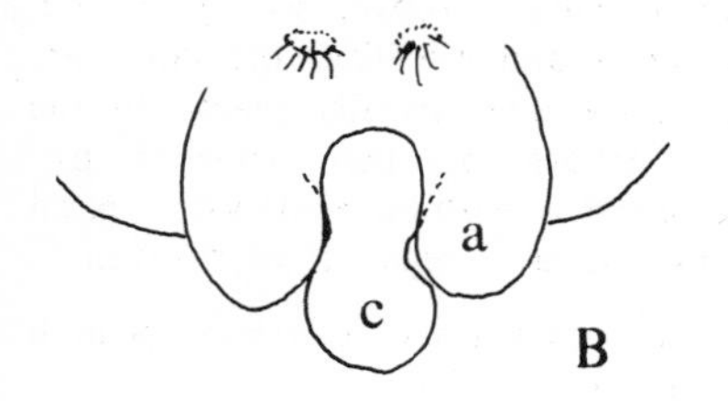

Fig. 66. Posterior part of abdomen of Drepanosiphidae, showing the rudimentary gonapophyses; ventral view. – A: *Drepanosiphum* (Drepanosiphinae) with 3 gonapophyses; B: *Thripsaphis* (Phyllaphidinae) with 2 gonapophyses; C: *Chaitophorus* (Chaitophorinae) with 4 gonapophyses. – a = anal plate; c = cauda.

2 (1) Compound eye with distinct triommatidion (ocular tubercle present) (Fig. 15 A). Empodial hairs flattened (Fig. 27 B). **Phyllaphidinae: Phyllaphidini**

– Triommatidion absorbed into the convexity of the compound eye (ocular tubercle absent). Empodial hairs flattened (Fig. 27 B) or simple (Fig. 27 A). .... **Phyllaphidinae: Saltusaphidini**

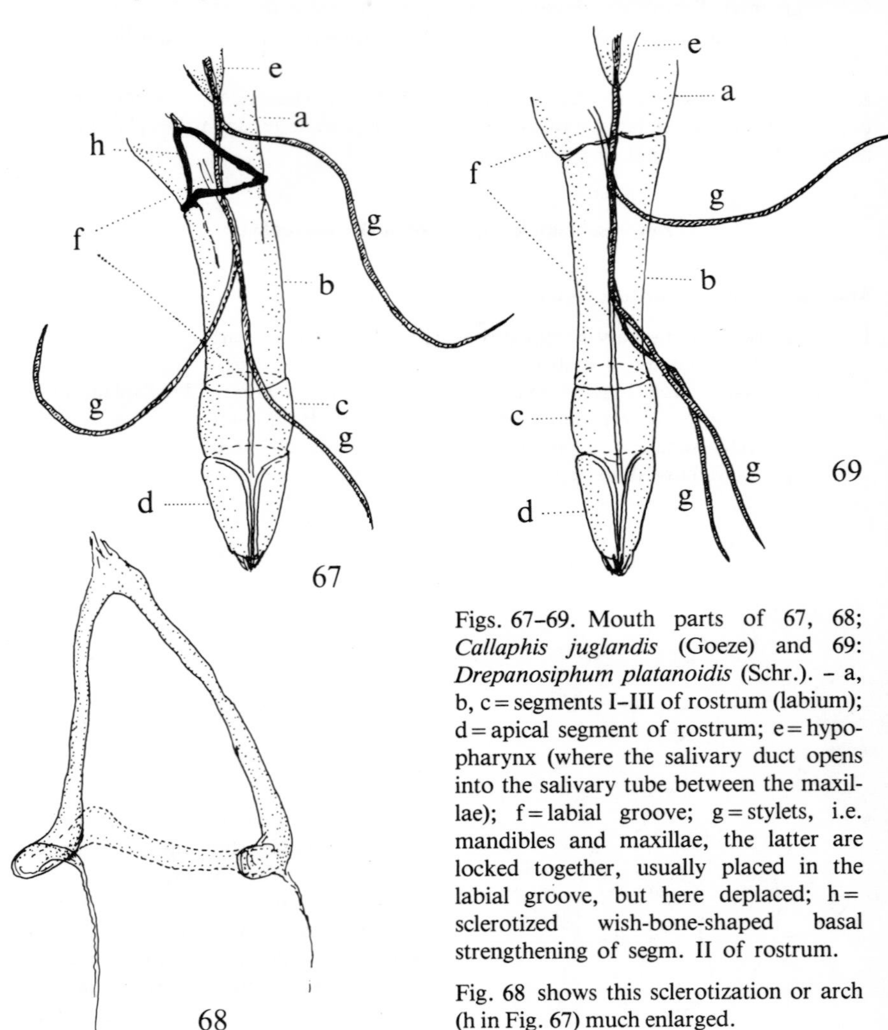

Figs. 67–69. Mouth parts of 67, 68; *Callaphis juglandis* (Goeze) and 69: *Drepanosiphum platanoidis* (Schr.). – a, b, c = segments I–III of rostrum (labium); d = apical segment of rostrum; e = hypopharynx (where the salivary duct opens into the salivary tube between the maxillae); f = labial groove; g = stylets, i.e. mandibles and maxillae, the latter are locked together, usually placed in the labial groove, but here deplaced; h = sclerotized wish-bone-shaped basal strengthening of segm. II of rostrum.

Fig. 68 shows this sclerotization or arch (h in Fig. 67) much enlarged.

3 (1) With three rudimentary gonapophyses (Fig. 66 A). Scandinavian species with subcylindrical siphunculi, which are 0.2 × body length or longer (Fig. 30 M). .... **Drepanosiphinae**

– With four rudimentary gonapophyses (Fig. 66 C). Siphunculi pore-shaped or stump-shaped, much shorter than 0.2 × body length (Fig. 30 L). .... **Chaitophorinae** – 4

4 (3) Antenna 6-segmented (the limit between segments III and IV may be indistinct). Siphunculi with at least some trace of reticulate sculpture, and more or less stump-shaped. **Chaitophorinae : Chaitophorini**

– Antenna 4- or 5-segmented. Siphunculi rim- or pore-like, truncate conical, or vasiform, without any trace of reticulate sculpture. .... **Chaitophorinae: Siphini**

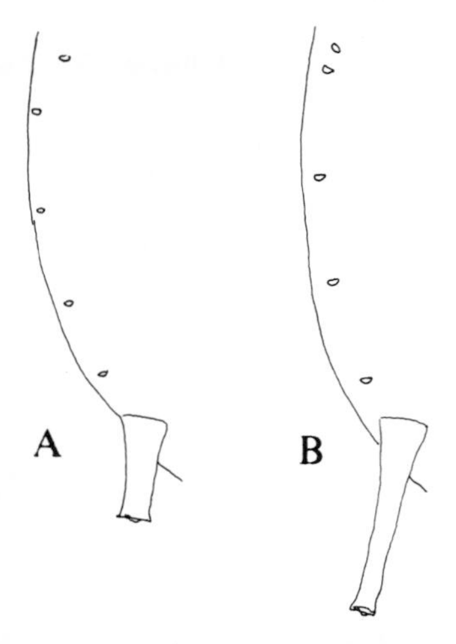

Fig. 70. Arrangement of the anterior abdominal stigmal pori (nos. 1–5) in A: *Aphis* (Aphidini) and B: *Uroleucon* (Macrosiphini).

Fig. 71. Hind leg of *Trama* (Lachnidae: Traminae). (After Szelegiewicz, redrawn).

Key to subfamilies and tribes of Aphididae

Apterous and viviparous females

1 Cauda semicircular, with more than 20 hairs. First segment of fore tarsus with 5 hairs. .... **Pterocommatinae**

– Cauda not semicircular, with less than 20 hairs, or, if cauda semicircular and with more than 20 hairs, then first segment of fore tarsus with at most 4 hairs. .... **Aphidinae** – 2

2 (1) Distance between stigmal pori on abdominal segments I and II about 3 × diameter of stigmal porus or longer, not shorter than 0.5 × distance between the stigmal pori on abdominal segments II and III (Fig. 70 A). Marginal tubercles on abdominal segments I and VII not smaller than those on II–V (which may be absent); or absent from all abdominal segments. **Aphidinae: Aphidini**

– Distance between stigmal pori on abdominal segments I and II shorter than 3 × diameter of stigmal porus, usually shorter than 0.5 × distance between stigmal pori on segments II and III (Fig. 70 B), or, if a little longer, then stigmal pori circular and very large. Marginal tubercles often present on abdominal segments II–V, but rarely on I and VII, and then the latter are smaller than those on II–V. .... **Aphidinae: Macrosiphini**

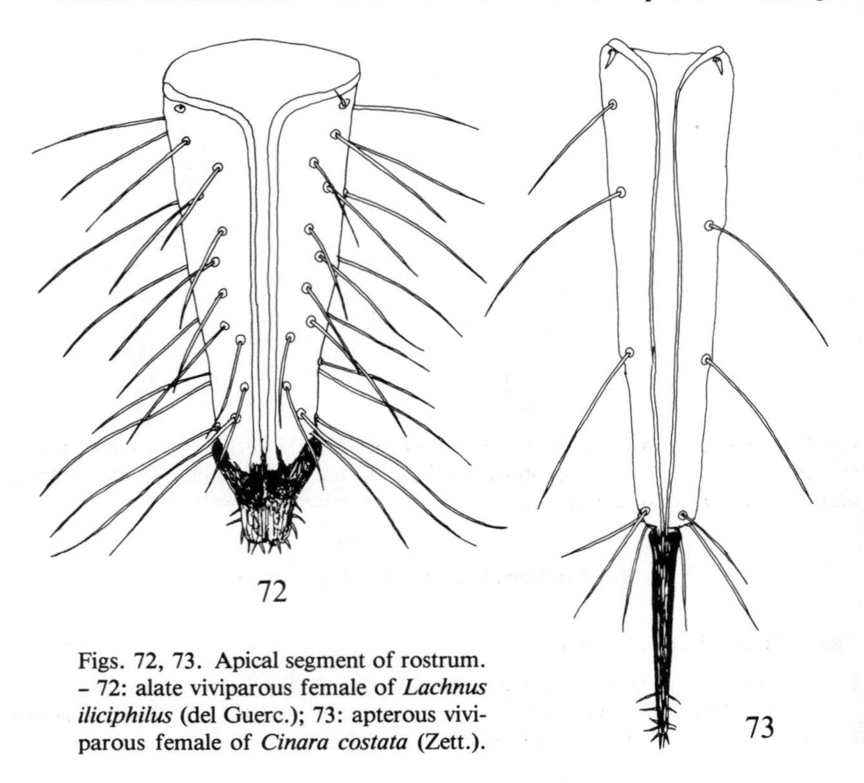

Figs. 72, 73. Apical segment of rostrum. – 72: alate viviparous female of *Lachnus iliciphilus* (del Guerc.); 73: apterous viviparous female of *Cinara costata* (Zett.).

Key to subfamilies and tribes of Lachnidae

Apterous and alate viviparous females

1 Hind tarsus about 0.5 × length of hind tibia or longer (Fig. 71) .... **Traminae**

– Hind tarsus distinctly shorter than 0.5 × length of hind tibia. 2

2 (1) Rostrum densely covered with hairs; ratio of number of accessory hairs on apical segment to its length in mm at least 35. Media of fore wing distinct, with two forks. Radial sector curved and rather long. Apical segment of rostrum obtuse and usually short (Fig. 72); if prolonged then rostrum about twice as long as body. Not on needles of *Pinus*. .... **Lachninae** – 3

– Rostrum with rather few hairs; ratio of number of accessory hairs on apical segment to its length in mm at most 30. Media of fore wing indistinct, unbranched, or with one or two forks. Radial sector straight and short. Apical segment of rostrum pointed and elongate (Fig. 73); if short and obtuse, then on needles of *Pinus*, and then media only exceptionally with two forks. .... **Cinarinae** – 4

3 (2) Rostrum about twice as long as body or longer. Processus terminalis with 20 or more hairs, all or most of which are similar to those on basal part of ultimate antennal segment (Fig. 74). .... **Lachninae: Stomaphidini**

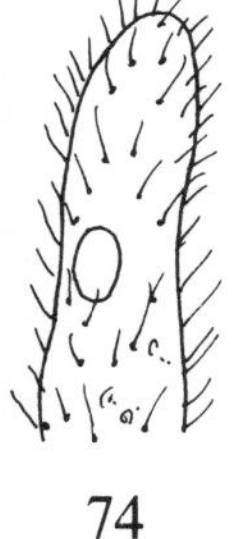

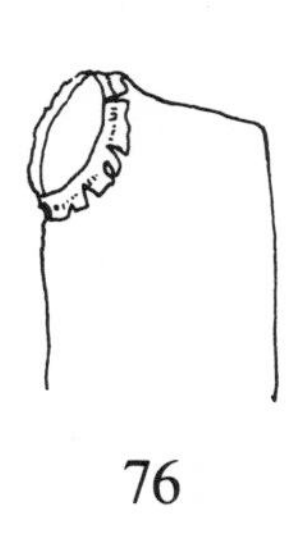

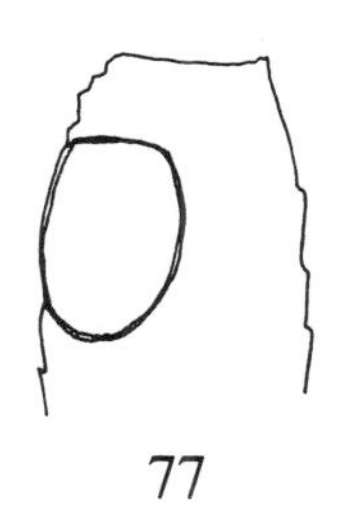

Figs. 74, 75. Apical part of ultimate antennal segment. – 74: *Stomaphis;* 75: *Lachnus*. (After Shaposhnikov, redrawn).

Figs. 76, 77. Apical part of penultimate antennal segment of 76: *Schizolachnus pineti* (F.), with small sclerites surrounding the primary rhinarium; 77: *Eulachnus agilis* (Kalt.), without such sclerites.

– Rostrum shorter, or at most a little longer, than body. Processus terminalis with at most 12 hairs, which are shorter than those on basal part of ultimate antennal segment (Fig. 75). **Lachninae: Lachnini**

4 (2) Apical segment of rostrum slender and pointed, 2–5 times as long as broad (Fig. 73). .... **Cinarinae: Cinarini**

– Apical segment of rostrum short, rather thick and obtuse, 0.6–1.2 times as long as broad. 5

5 (4) Body at most 1.8 × as long as wide. Primary rhinaria with rings of small sclerites (Fig. 76). First segment of hind tarsus shorter than 0.4 × second segment (Fig. 78). Triommatidia present. .... **Cinarinae: Schizolachnini**

– Body at least 2 × as long as wide. Primary rhinaria without rings of small sclerites (Fig. 77). First segment of hind tarsus longer than 0.5 × second segment (Fig. 79). Triommatidia absent. .... **Cinarinae: Eulachnini**

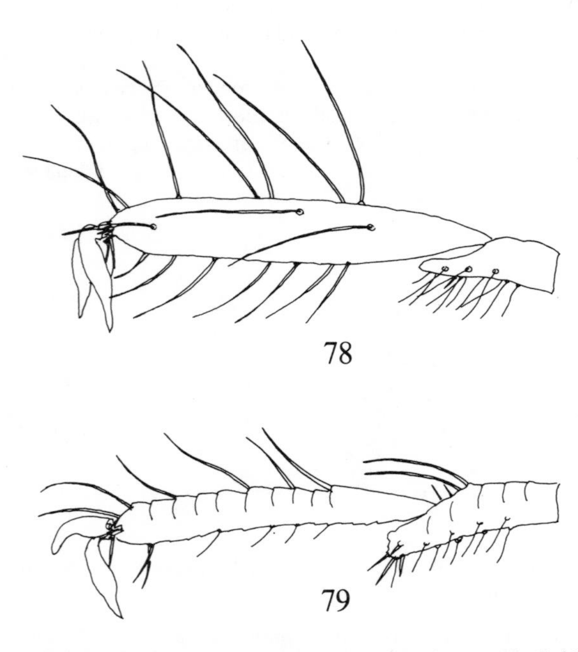

Figs. 78, 79. Hind tarsi of apterous viviparous females – 78: *Schizolachnus pineti* (F.); 79: *Eulachnus rileyi* (Will.).

Family Mindaridae

Eyes of apterous individuals and young nymphs with only three ommatidia and without distinct limit between head and pronotum. Wax glands well developed. Siphunculi pore-shaped. Cauda subtriangular. Alate individuals with transverse oval secondary rhinaria. Pterostigma narrow, pointed, prolonged to wing apex; radial sector originating from base of pterostigma; wings roof-like in repose. Sexuales apterous, dwarfish, but with well developed rostrum.

Only one genus.

Genus *Mindarus* Koch, 1857

Mindarus Koch, 1857: 277.

Type-species: *Mindarus abietinus* Koch, 1857.

Survey: 284.

Abdomen with marginal facetted wax gland plates (Fig. 85); the posterior segments also with dorsal plates. Frons convex, without tubercles. Antennae 6-segmented. Processus terminalis 0.17–0.25 × VIa. Apterae without secondary rhinaria, alatae with transverse oval, rather narrow secondary rhinaria on ant. segm. III; segm. IV with 1–4 rhinaria close to distal end. Media of fore wing usually with one fork. Radial sector starts at the basal part of the very long, narrow, curved, pointed pterostigma, which continues to apex of wing. Hind wing with two oblique veins. Siphuncular pores present. Cauda short, tongue-shaped subtriangular (Fig. 84). Anal plate not bilobed.

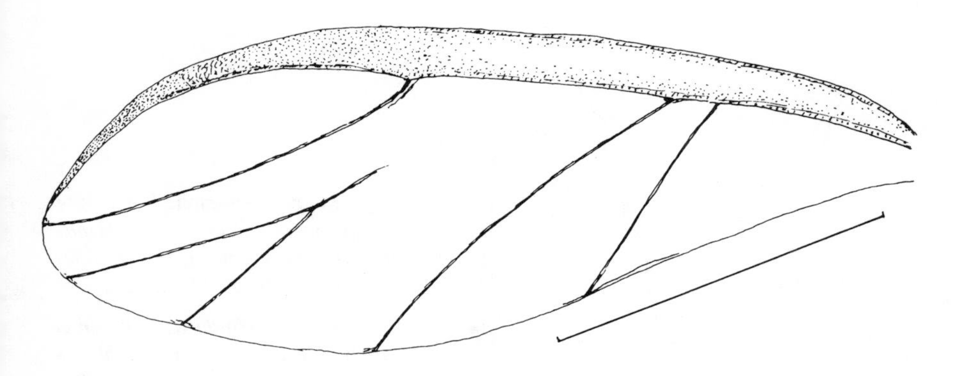

Fig. 80. Fore wing of *Mindarus obliquus* (Chol.). (Scale 1 mm).

Holarctic. The genus is known from the early Tertiary. Nine species are described from Europe, Asia, and N America, four being recent. They feed on conifers (*Abies, Picea*). Two species in Scandinavia.

Key to species of *Mindarus*

Apterous viviparous females

1 On *Abies*. .... 1. *abietinus* Koch
– On *Picea*. .... 2. *obliquus* (Cholodkovsky)

Alate viviparous females

1 Ant. segm. III with (12–) 14–27 rhinaria, usually 0.31 mm or longer. Ant. segm. IV usually with only one rhinarium (Fig. 82). .... 1. *abietinus* Koch
– Ant. segm. III with 7–12 (–14) rhinaria, usually 0.31 mm or shorter. Ant. segm. IV usually with 2–4 rhinaria (Fig. 81). 2. *obliquus* (Cholodkovsky)

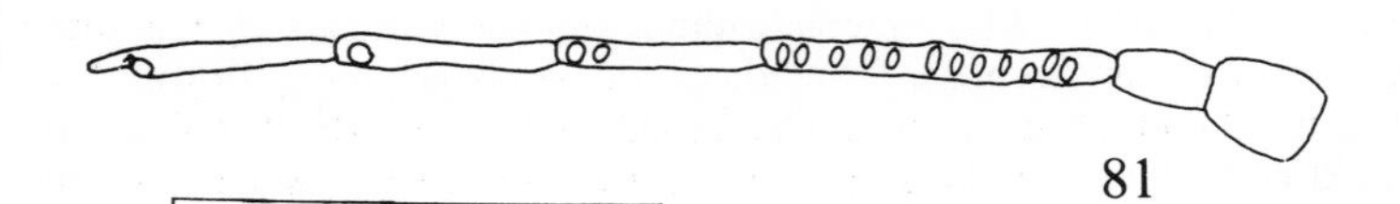

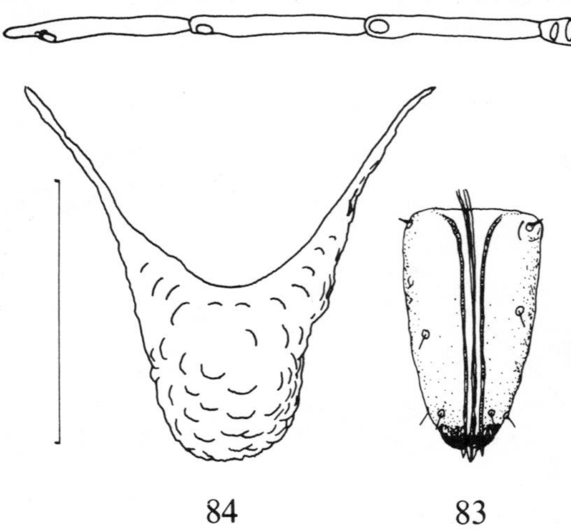

Figs. 81, 82. Antennae of alate viviparous females. – 81. *Mindarus obliquus* (Chol.); 82: *M. abietinus* Koch. (Scale 0.5 mm).

Figs. 83, 84. *Mindarus abietinus* Koch, alate viviparous female. – 83: apical segment of rostrum; 84: cauda. (Scale 0.1 mm).

1. ***Mindarus abietinus*** Koch, 1857
Figs. 82–84.

Mindarus abietinus Koch, 1857: 278. – Survey: 284.

Apterous viviparous female (including fundatrix). Yellowish green, covered with wax wool. Antennae and legs slightly darker than body. Antennae shorter than 0.5 × body. Rostrum reaching past hind coxae; apical segm. about 0.5 × 2sht., blunt. 1.7–2.0 mm.

Alate viviparous female. Head and thorax dark. Abdomen with dark transverse stripes and marginal spots. Antennae 0.5 × body or longer; segm. III with 12–27 secondary rhinaria spread over entire segment (Denmark: 14–22, Finland: 14–27, C Europe: 12–19), IV with 1–3 (usually only one) in the distal part. Ant. segm. III 0.30–0.50 mm, usually longer than 0.31 mm. Apical segm. of rostrum 0.4 × 2sht.

Sexuales. Dwarfish, apterous.

Distribution. In Denmark rather common in Jutland, probably all over the country; in Sweden from Sk. north to Dlr.; known from trap in AK, Norway; in Finland from Ab and N north to ObS. – Widespread in Europe, including all countries bordering the Baltic and the North Sea, but rare in N Germany and Britain; in Asia known from W Siberia and the Middle East (Turkey, Lebanon); originally palaearctic, but introduced into N America and now widespread in the USA and Canada.

Biology. The species lives between the needles on young shoots of Silver Fir (*Abies alba, A. nordmanniana;* occasionally also on other *Abies* spp.). There are only three generations a year: 1) apterous fundatrices, 2) alate and apterous viviparous females (sexuparae), and 3) sexuales. The species is a pest to *Abies*. The needles may become shortened, pale, and twisted, the shoots may lose the needles and get bark crevices. The top shoot may be killed and replaced by side branches, so that the value of the trunk is diminished. The small sexuales are born in early summer or in mid summer so that the species is represented only by eggs most of the year, from summer to next spring. The eggs are black, covered with small, white bits of wax threads so that they look greyish (Fig. 41).

2. ***Mindarus obliquus*** (Cholodkovsky, 1896)
.Plate 1: 2–3. Figs. 80, 81, 85–91.

Schizoneura obliqua Cholodkovsky, 1896: 257. – Survey: 284.

Apterous viviparous female (including fundatrices). Very similar to *abietinus*.

Alate viviparous female. Similar to *abietinus,* but with fewer secondary rhinaria on ant. segm. III, 7–13 (usually 7–12, occasionally with 14 rhinaria on one antenna), and with more secondary rhinaria on IV, 1–4, usually 2–3, located near the distal end. Ant. segm. III on an average shorter than in *abietinus,* 0.22–0.36 mm, usually shorter than 0.31 mm.

Oviparous female. With two ventral, pigmented, reticulate wax gland plates on the posterior part of abdomen. Rostrum reaches to middle of abdomen. Hind tibia hardly thickened, with 5–6 indistinct scent plaques of irregular shape. About 1 mm.

Distribution. In Denmark widespread, more common than *abietinus;* in Sweden found in Sk.; known also from Norway; in Finland found in Sb. – Widespread in Europe, but rare in N Germany, apparently rare in Poland (not at the Baltic coast), unknown to Britain before 1967; seems to be common in C Europe and N Russia; also recorded from Turkey.

Biology. The species feeds on various species of *Picea,* e. g. *P. glauca* and *P. sitchensis* (but not *P. abies*). The biology is similar to *abietinus,* the aphids occurring from spring to midsummer between the needles of the young shoots. The needles do not become twisted. The colonies are easy to detect because of the cover of cotton-like wax.

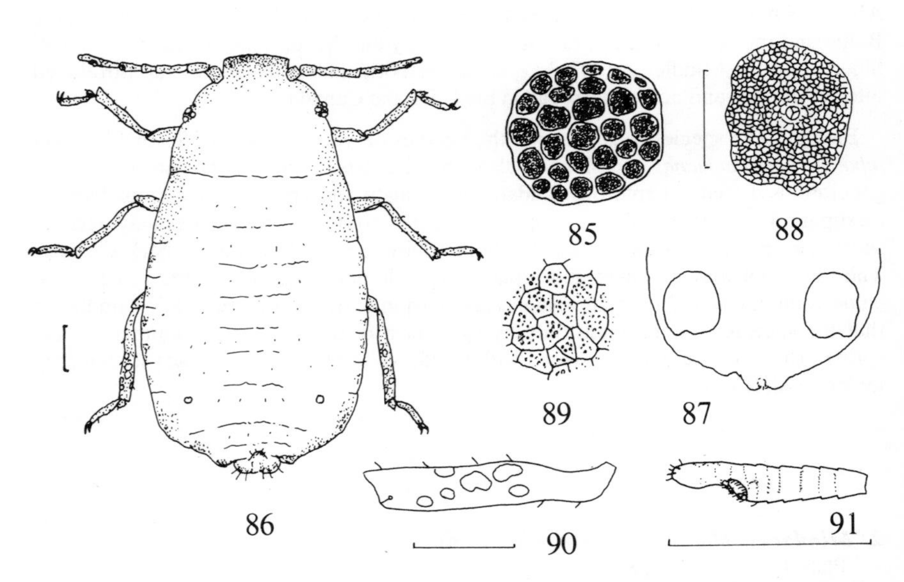

Figs. 85–91. *Mindarus obliquus* (Chol.). – 85: wax gland of alatoid nymph; 86: oviparous female; 87: posterior part of abdomen of oviparous female in ventral view showing position of ventral wax gland plates; 88: ventral wax gland plate of oviparous female; 89: sculpture of small part of this plate on a larger scale; 90: hind tibia of oviparous female: 91: ant. segm. VI of oviparous female. (Scales: 0.1 mm).

Family Hormaphididae

Eyes of apterous individuals and young nymphs with only three ommatidia and without distinct limit between head and pronotum. Antennae with 3–5 segments, much shorter than body. Antennae of alatae with narrow, ringlike secondary rhinaria. Cubitus 1a and 1b of fore wing leave the main vein at the same point and sometimes form a common stalk. Wings flat in repose. Tarsi with two long, capitate dorsoapical hairs (Fig. 96). Apterae and nymphs often coccid- or aleyrodid-like and more or less sedentary, with very short legs concealed beneath the body. Wax production from special pores. Siphunculi pore-shaped or absent. Cauda knobbed. Anal plate bilobed (Fig. 60). Sexuales apterous, with well developed rostrum (sexuales of Scandinavian species have not been found).

Many species in the Tropics and Subtropics, only few in Europe. Some species are dioecious with a holocycle lasting two years and produce galls on the primary hosts. All Scandinavian species are anholocyclic.

Key to subfamilies and genera of Hormaphididae

Apterous viviparous females

1 Frons with two horn-like processes (Fig. 92 C). Siphuncular pores present. (Oregminae) .... *Cerataphis* Lichtenstein (p. 88)
– Frons without horn-like processes. Siphuncular pores absent. (Hormaphidinae) 2
2 (1) Body surrounded by a fringe of wax threads produced by marginal glands (Fig. 98 A). .... *Hormaphis* Osten-Sacken (p. 91)
– Body wax-powdered, but not surrounded by a wax fringe. *Hamamelistes* Shimer (p. 89)

Alate viviparous females

1 Media of fore wing with one fork (Fig. 92 D). (Oregminae). *Cerataphis* Lichtenstein (p. 88)
– Media of fore wing unbranched (Fig. 93). (Hormaphidinae) 2
2 (1) Hind wing with two oblique veins (Fig. 93). .... *Hamamelistes* Shimer (p. 89)
– Hind wing with only one oblique vein (sometimes with a fork, Fig. 98 C). .... *Hormaphis* Osten-Sacken (p. 91)

SUBFAMILY OREGMINAE

Frons of apterae with two horn-like processes.

Genus *Cerataphis* Lichtenstein, 1882

Cerataphis Lichtenstein, 1882: 75.
Type-species: *Coccus lataniae* Boisduval, 1867.
Survey: 129.

One species in the region, seven species in the world.

3. ***Cerataphis orchidearum*** (Westwood, 1879)
Fig. 92.

Asterolecanium orchidearum Westwood, 1879: 796. – Survey: 130.

Apterous viviparous female. Dark reddish brown or black, surrounded by a fringe of wax, dorsum powdered. Body flat and circular, with crenulated margins. Frons with two pointed processes or horns. With rather long, fine hairs on head and body, also around the siphuncular pores, longest on tergite VIII, cauda, and anal plate. Antennae

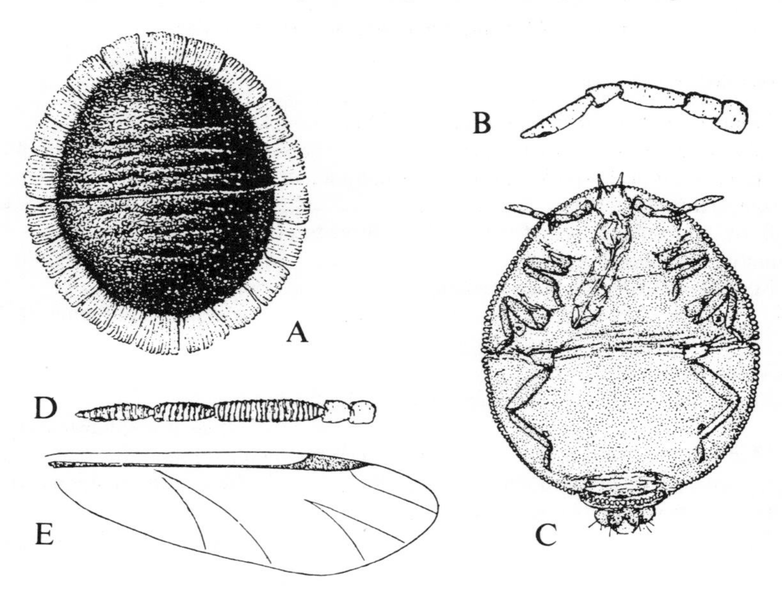

Fig. 92. *Cerataphis orchidearum* (Westw.). – A: apterous viviparous female in dorsal view, surrounded by a fringe of wax threads; B: antenna of aptera; C: apterous female in ventral view; D, E: antenna and fore wing of alate viviparous female. (After Börner & Heinze).

about 0.15 × body, 4- or 5-segmented. Rostrum reaches to middle coxae. Legs very short, but tarsi not reduced. Siphuncular pores distinct, placed on low cones. About 1.5 mm.

Alate viviparous female. Abdomen yellowish. Frons without horns. Antennae about 0.4 × body, 5-segmented; segm. III about as long as IV + V or a little longer, with about 20 ringlike secondary rhinaria, IV and V each with about 10. Legs longer than in apterae. Media of fore wing with one fork. Hind wing with two oblique veins. Siphuncular pores present. Cauda with about 10 hairs. 1.6–2.3 mm.

Distribution. Recorded from glasshouses in Sweden (Sk.) and Finland (N); not found in Denmark or Norway. – The origin is probably tropical Asia. Widespread in the Tropics, also of the New World. Introduced with cultivated orchids to subtropical and temperate regions.

Biology. The species is anholocyclic and lives on Orchidaceae, in Scandinavia only indoors. The apterae become sedentary before the final moult. Alatae are fairly rare.

SUBFAMILY HORMAPHIDINAE

Frons without horn-like processes.

Genus *Hamamelistes* Shimer, 1867

Hamamelistes Shimer, 1867: 284.
Type-species: *Hamamelistes spinosus* Shimer, 1867.
Survey: 212.

One species in the region, two species in the world.

4. ***Hamamelistes betulinus*** (Horvath, 1896)
Figs. 60, 93–97.

Tetraphis betulina Horvath, 1896: 6.
Hamamelistes tullgreni de Meijere, 1912: 93.
Survey: 212.

Apterous viviparous female. Dark brown, almost black, with white wax powder. Body oval. Antennae 3-segmented, very short, without secondary rhinaria. Processus terminalis very indistinct. Legs more or less reduced, especially in the generation overwintering as young nymphs; the tarsi of these are missing on fore and middle legs and extremely small on hind legs. Siphuncular pores absent. About 1.5 mm.

Alate viviparous female. Head and thorax brown, abdomen green or greenish brown. Antennae 5-segmented, about 0.33 × body; segm. III with 22–25 ringlike secondary rhinaria, IV with 8–12, V with 8–11. Rostrum short, not reaching middle coxae; apical segm. 0.7–0.9 × 2sht. Fore wing with unbranched media. Hind wing with two oblique veins leaving the main vein at separate points. Siphuncular pores present (even from birth). 1.3–2.0 mm.

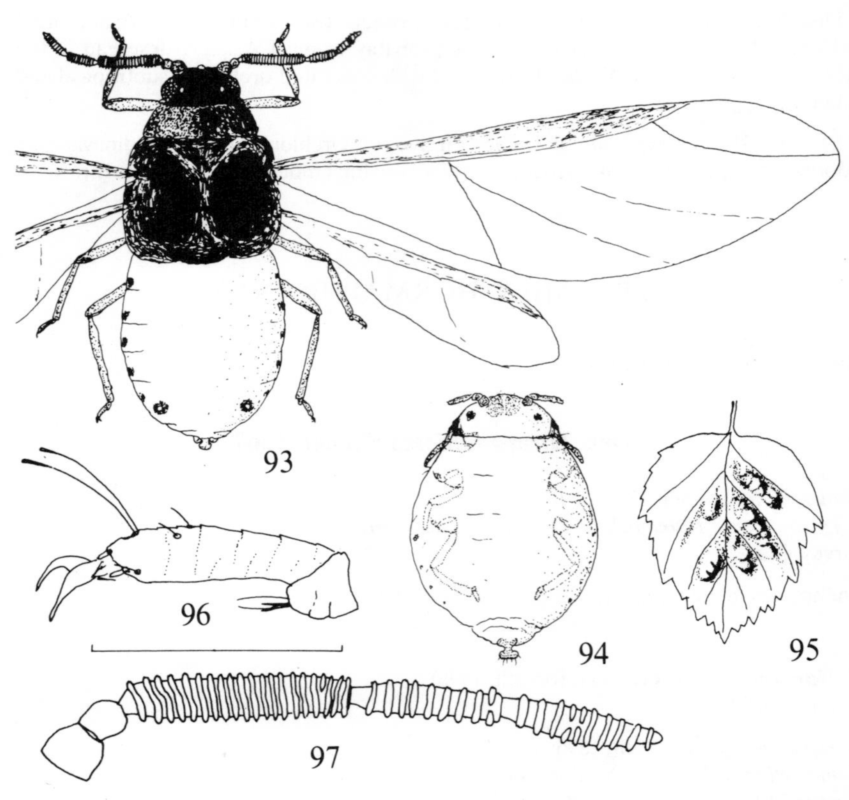

Figs. 93–97. *Hamamelistes betulinus* (Horv.). – 93: alate viviparous female (from Denmark); 94: apterous viviparous female in dorsal view (legs in dotted lines, only visible from the underside; rostrum not drawn); 95: birch leaf with galls, upper surface; 96: hind tarsus of alate viviparous female; 97: antenna of alate viviparous female. (Scale: 1 mm for 93 and 94, 0.1 mm for 96). (95 redrawn after Mordvilko, others original).

Distribution. In Denmark found in several localities in Jutland and Zealand, but not common; in Sweden known in several districts from Sk. north to L. Lpm.; no records from Norway; in Finland found in N, Oa, and ObN. - Widespread in W, N, C & E Europe, but not common, in Germany rare; recorded from Britain, Poland, and N Russia, but not from N Germany; in Asia known from W Siberia and Mongolia.

Biology. It lives on birch *(Betula pubescens, B. verrucosa)* and is anholocyclic. Coccid-like 1st instar nymphs (the hiemalis generation) hibernate on young branches and become apterous adults when the birch buds open. Their offspring (aestivalis) move to the leaves and produce low yellowish green blisters, paler than the rest of the leaf. The colonies feed on the concave undersides of the blisters and are usually found on rather few leaves of the individual tree. This generation is apterous according to observations in England and other countries, where alate viviparous females are said to occur only in the second aestivalis-generation. However, own observations in Denmark of alatae and alatoid nymphs in small colonies as early as in the beginning of June show that they may occur also in the first aestivalis-generation. The colonies may be visited by ants. Hiemalis-nymphs are born by alatae and apterae of 1st and 2nd generations. The subspecies *miyabei* Matsumura, which lives in Japan, is holocyclic and host-alternating.

Note. This is the species called *Hamamelistes betulae* (Mordv.) by Tullgren (1909), who found it in Stockholm. Alatae occurred here from the middle of July to the middle of August. De Meijere (1912) named it *H. tullgreni* because he believed it was different not only from *H. betulae,* but also from *H. betulinus* (Horv.).

Genus *Hormaphis* Osten-Sacken, 1861

Hormaphis Osten-Sacken, 1861: 422.
Type-species: *Hormaphis hamamelidis* Osten-Sacken, 1861.
= *Byrsocrypta hamamelidis* Fitch, 1851.
Survey: 216.

One species in the region, four species in the world.

5. ***Hormaphis betulae*** (Mordvilko, 1901)
Figs. 98, 99.

Cerataphis betulae Mordvilko, 1901: 973. - Survey: 217.

Apterous viviparous female. Yellow, greenish, or yellowish brown. Body flat and circular, surrounded by a fringe of beam-like wax threads. Antennae short and 3-segmented. Legs strongly reduced. Siphuncular pores absent. About 1 mm.

Alate viviparous female. Very similar to *Hamamelistes betulinus,* but hind wing with only one oblique vein (or two veins having a common stalk), and siphuncular pores absent. Ant. segm. III with 21 secondary rhinaria, IV with 11, V with 12 (according to de Meijere (1912)).

Distribution. In Scandinavia recorded only from Finland (district N). – Widespread in Europe, especially in and north of the Alps, in Poland, Hungary, and W & NW Russia, but only locally common. Unknown from the Baltic region of Poland and from N Germany. Also Siberia.

Biology. The species is anholocyclic and feeds on *Betula.* The biology is poorly known, but seems to correspond well to *Hamamelistes betulinus.* Small nymphs hibernate in the soil. In spring they climb the trunks and enter the lower leaves of the young

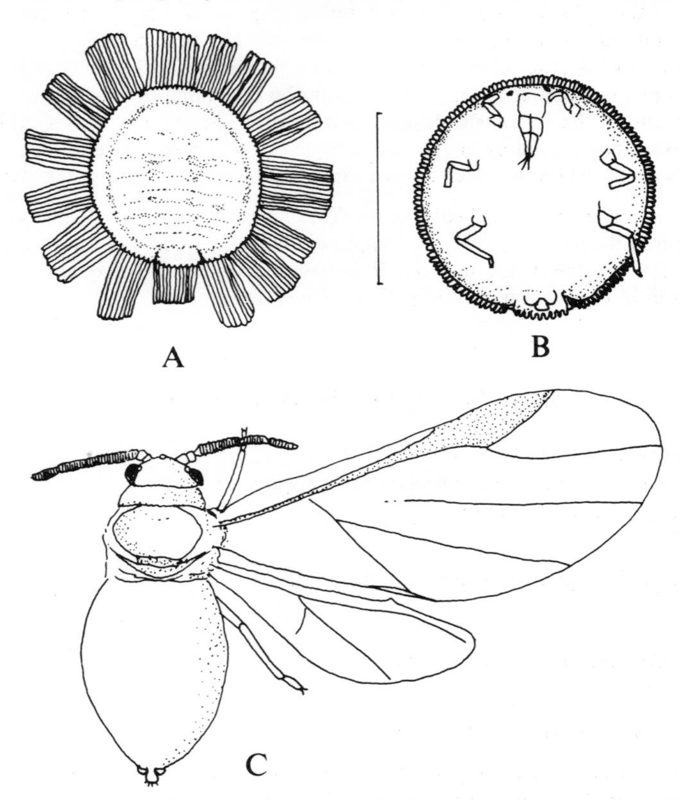

Fig. 98. *Hormaphis betulae* (Mordv.). – A, B: apterous viviparous female in dorsal (A) and ventral (B) view; the wax threads are drawn in A, but not in B; C: alate viviparous female. (After Mordvilko, redrawn). (Scale 1 mm).

shoots. They become apterous adults as early as in the 4th instar. The leaves are not deformed. Alate viviparous females occur in the summer months. In autumn young nymphs fall to the ground with the leaves. The species apparently prefers young birches on sandy soil.

Note. Tullgren (1909) recorded *Hamamelistes betulae* (Mordv.) from Sweden, but the species was *Hamamelistes betulinus* (Horv.), which he regarded as a possible synonym.

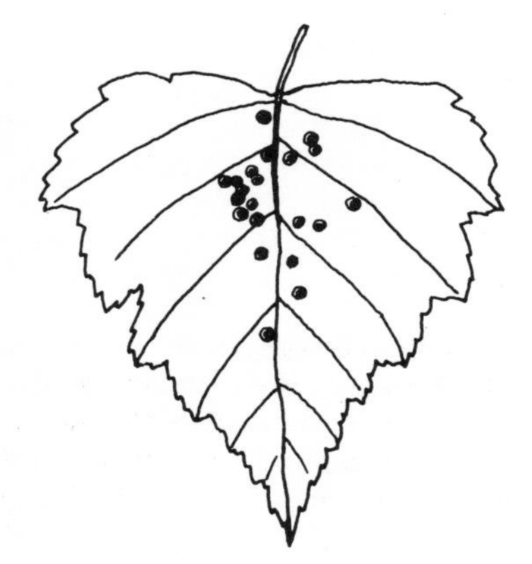

Fig. 99. Apterous *Hormaphis betulae* (Mordv.) on upper su:face of birch leaf *(Betula verrucosa)*. (Based on photo in Mordvilko 1935).

Family Thelaxidae

Eyes of apterae and young nymphs with only three ommatidia and without a distinct limit between head and pronotum. Antennae 5-segmented, with transverse rows of fine spinules. Processus terminalis much shorter than basal part of segm. V. Antennae of alatae with circular or subcircular secondary rhinaria. Apical segment of rostrum subdivided; ultimate part rather slender or needle-like (Figs. 102, 107). Media of fore wing with one fork. Hind wing with only one oblique vein. Wings flat in repose. Siphunculi pore-shaped or low and truncate. Cauda broadly rounded or knobbed. Anal plate rounded. Sexuales apterous, with well developed rostrum.

Key to genera of Thelaxidae

Apterous and alate viviparous females

1 Cauda knobbed (Fig. 100). First tarsal segments with 7 hairs.
Thelaxes Westwood (p. 94)

– Cauda semicircular (Fig. 101). First tarsal segments with 5 hairs.
Glyphina Koch (p. 95)

Genus *Thelaxes* Westwood, 1840

Thelaxes Westwood, 1840: 118.
Type-species: *Thelaxes quercicola* Westwood, 1840.
= *Aphis dryophila* Schrank, 1801.
Survey: 422.

One species in Scandinavia, three species in the world.

6. ***Thelaxes dryophila*** (Schrank, 1801)
Plate 1: 1. Figs. 30 P, 100, 102–104.

Aphis dryophila Schrank, 1801: 113. – Survey: 422.

Apterous viviparous female. Dark brownish red with yellowish green, longitudinal dorsal stripe. Antennae, legs, siphunculi, and cauda brownish. Body with two kinds of hairs, fine and spine-like, the latter especially distinct on posterior part of abdomen. Small marginal tubercles present on abd. segments. Antennae 5-segmented, shorter than 0.5 × body, with transverse rows (or rings) of fine spinules; these most distinct on distal segments; secondary rhinaria absent. Processus terminalis shorter than 0.5 × Va. Longest hair on ant. segm. III about 2.5 × IIIbd. Rostrum reaching hind coxae; apical segm. 1.5 × 2sht. or longer, slender, subdivided into two segments, the distal one being needle-shaped. First tarsal segments with 7-7-7 hairs. Hind tibia may have 1–6 scent plaques. Siphuncular pores rather large, placed on low cones (Fig. 30 P). Cauda knobbed, about as long as hind tarsi. Anal plate rounded. 1.1–2.3 mm.

Fundatrix. Relatively dark. Antennae 0.25–0.33 × body; processus terminalis 0.15–0.20 × Va. Hind tibia without scent plaques. 2.2–2.5 mm. Otherwise like the apterous viviparous female described above.

Alate viviparous female. Head and thorax black, abdomen with dark marginal spots, the posterior segments also with dark dorsal cross bands. Antennae, legs, cauda, and areas around siphuncular pores dark. Ant. segm. III with 4–7 rather large, circular secondary rhinaria. Media of fore wing with one fork. Hind wing with only one oblique vein. Wings flat in repose.

Distribution. Common and widespread in Denmark and S Fennoscandia; in Sweden north to Upl.; in Norway north to HOy; in Finland north to Ta. – In Europe from Britain in the West to Russia, Caucasia, and Kazachstan in the East, from Scandinavia in the North to Portugal, Spain, and Turkey in the South; common in N Germany and N Poland; introduced in N America.

Biology. Oak *(Quercus)* is the only host. In Denmark fundatrices become adults in May. After that time colonies are found on tips of shoots, young stems, petioles, and undersides of leaves, especially along the mid ribs. Colonies have not been observed in

Denmark after August. Small nymphs, which may be aestivating morphs, have been found in England in August. Sexuales are apparently undescribed.

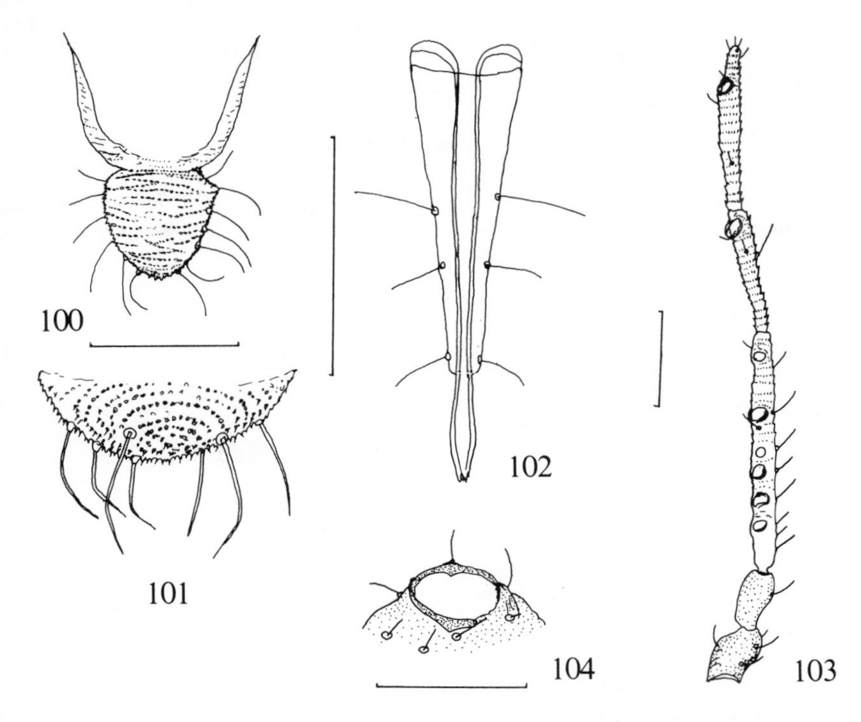

Figs. 100, 101. Cauda of apterous viviparous females. – 100: *Thelaxes dryophila* (Schr.); 101: *Glyphina betulae* (L.). (Scale 0.1 mm).

Figs. 102–104. *Thelaxes dryophila* (Schr.). – 102: apical segment of rostrum of apt. viv.; 103: antenna of al. viv.; 104: siphuncular pore of al. viv. placed on haired, dark cone. (Scales 0.1 mm).

Genus *Glyphina* Koch, 1856

Glyphina Koch, 1856: 259.
Type-species: *Vacuna betulae* Kaltenbach, 1843.
= *Aphis betulae* Linné, 1758.
Survey: 207.

First tarsal segments with 5-5-5 hairs. Siphunculi low, truncate, placed on low warts. Cauda semicular. Otherwise very similar to *Thelaxes*.

Five species in the world. Two species are recorded from Scandinavia, but they seem to be different only with regard to host plant affinity, and not morphologically. Consequently they may well be regarded as subspecies or even synonyms. A morphological key cannot be given.

Key to species of *Glyphina*

1 On *Betula* .... 7. *betulae* (Linné)
– On *Alnus* .... 8. *schrankiana* Börner

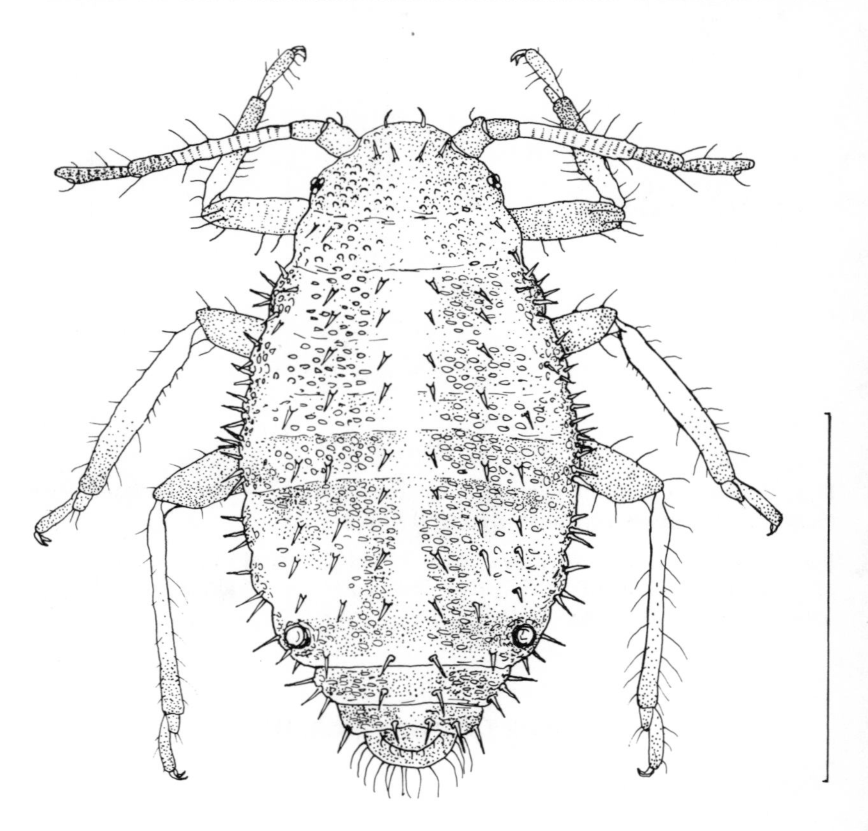

Fig. 105. *Glyphina betulae* (L.), apt. viv. (Scale 1 mm).

7. ***Glyphina betulae*** (Linné, 1758)
Figs. 101, 105–110.

Aphis betulae Linné, 1758: 452. – Survey: 207.

Apterous viviparous female. Dark green or black (as young nymph green), with white longitudinal, dorsal stripe and fine intersegmental, transverse stripes. Antennae (except segm. III), legs, and siphunculi dark. Body hairs strong, thornlike; the marginal hairs on abd. segm. IV about 1.5 × IIIbd. Cuticle densely covered with distinct small warts. Antennae shorter than 0.5 × body, without secondary rhinaria; longest hair on segm. III 1.7–2 × IIIbd. Rostrum reaching hind coxae; apical segm. 1.1–1.4 × 2sht. 1.5–2.1 mm.

Alate viviparous female. Head and thorax black; abdomen with marginal spots and transverse, dorsal rows of small, dark spots. Ant. segm. III with 2–7 almost circular, very small secondary rhinaria.

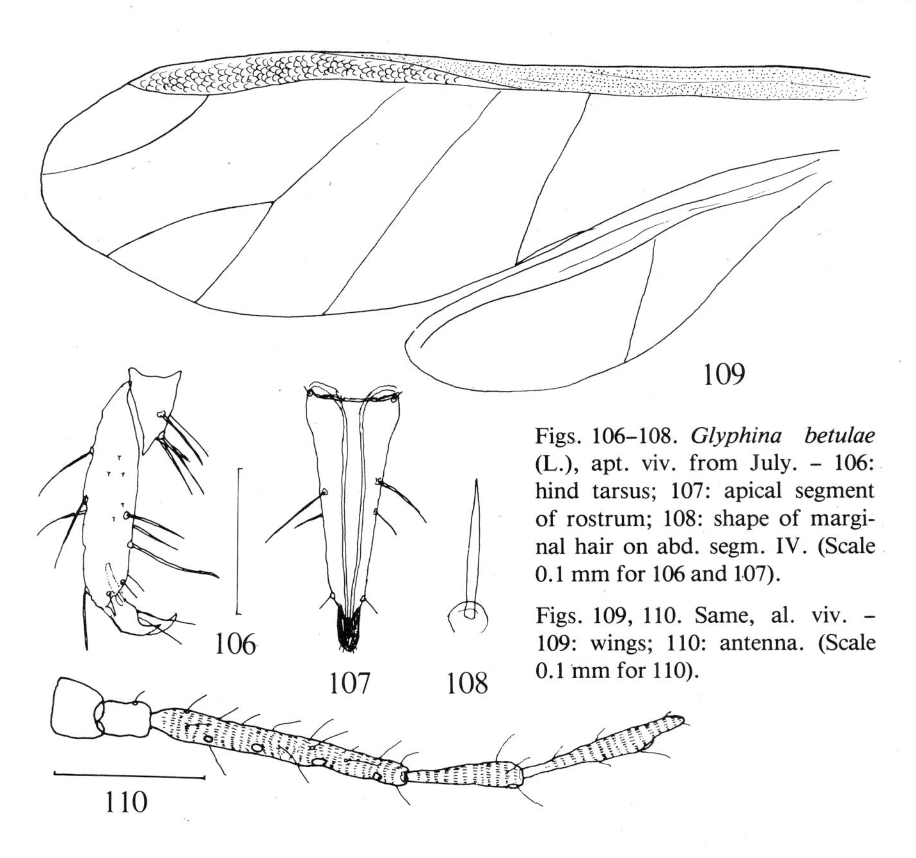

Figs. 106–108. *Glyphina betulae* (L.), apt. viv. from July. – 106: hind tarsus; 107: apical segment of rostrum; 108: shape of marginal hair on abd. segm. IV. (Scale 0.1 mm for 106 and 107).

Figs. 109, 110. Same, al. viv. – 109: wings; 110: antenna. (Scale 0.1 mm for 110).

Oviparous female. Hind tibia not swollen, with 0–6 scent plaques. Otherwise rather similar to the apterous viviparous female.

Male. Apterous. Almost black. Antennae without secondary rhinaria. About 1 mm.

Distribution. Rather common and widespread in Denmark; in Sweden north to P.Lpm.; known from Norway; in Finland north to ObS. – Probably originally palaearctic, recorded from Great Britain, N Germany (common), Poland (including the Baltic coast region), C Europe, Russia (also in the Baltic region), and W Siberia; introduced in N America and now widespread in the USA and Canada.

Biology. The only host is *Betula*. The species is recorded from both *B. verrucosa* and *B. pubescens*, but in Denmark only from *B. pubescens*. The life cycle is abbreviated; sexuales appear rather early, in the middle of the summer. I have observed copulating oviparous females and males in August, in 1962 also in early September, but the colonies usually disappears long before. They are found on young shoots in spring and summer, and are visited by ants. Most individuals are apterous viviparous females. Alate viviparous females are produced in June, July and August, but primarily in early summer.

8. ***Glyphina schrankiana*** Börner, 1950

Glyphina schrankiana Börner, 1950: 17. – Survey: 207.

Börner (1950) gave the following uncertain morphological differences between *schrankiana* and *betulae:* dorsal hairs longer and marginal sclerites larger, touching the spinopleural sclerites and – on abdomen – fusing with these.

Distribution. Not in Denmark; Danielsson (1979, in litt.) has a balsam slide of a *Glyphina* collected by Tullgren on *Alnus* in Sweden (Upl., Stockholm 1904); in Norway recorded from HOi; in Finland recorded from N, Ta, Sa, and Kb. – Known from several localities in Europe, e.g. in Poland and Russia (also in the Baltic regions of both countries), but not in N Germany; outside Europe found in the Caucasus region and – according to Archibald (1957: 39) in Canada (Nova Scotia).

Biology. The biology is similar to *betulae*, but the host is *Alnus* instead of *Betula*.

Note. Aphids with an abbreviated life cycle and only a small production of alatae may easily evolve local populations with different gene pools. It is open for discussion how different such populations shall be to be regarded as two species. Further studies are needed to show if *schrankiana* is conspecific with *betulae* or if they represent two good species restricted to *Alnus* and *Betula*.

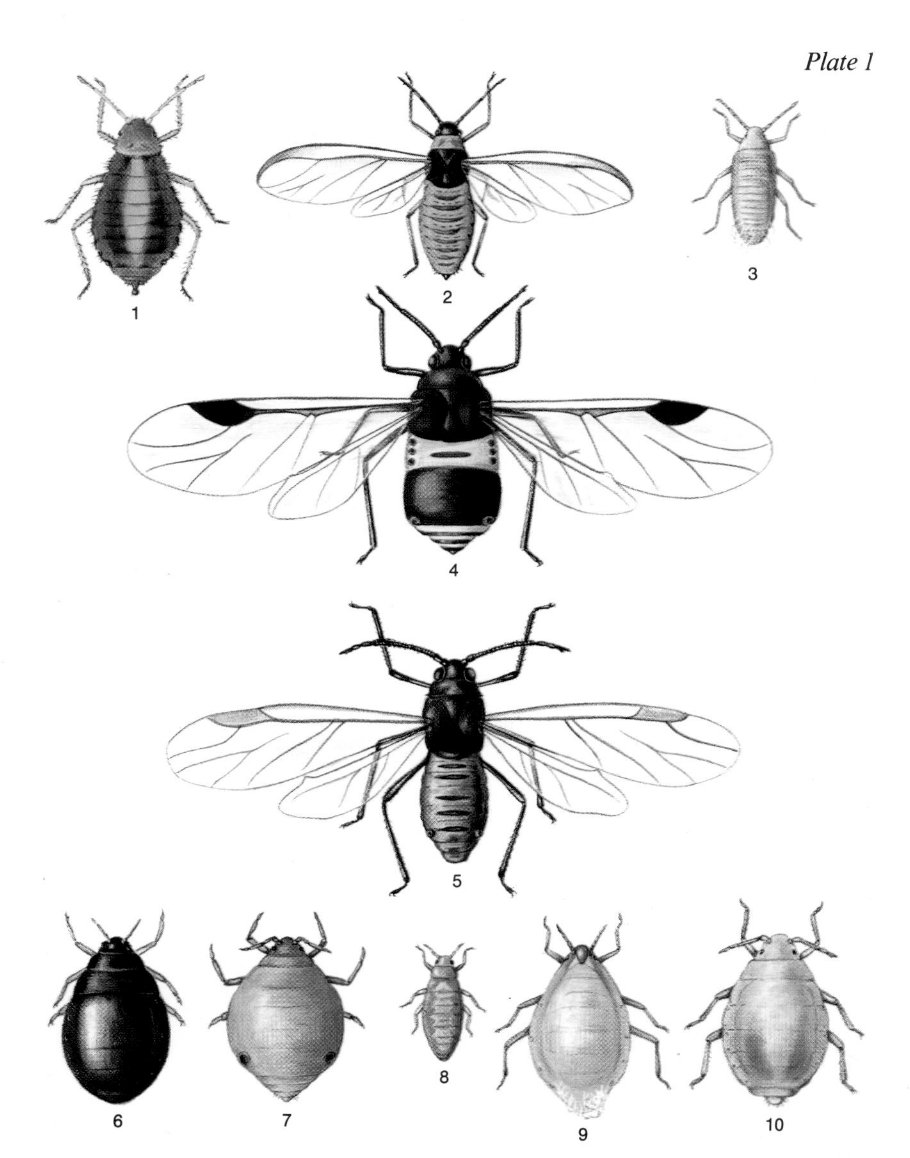

1. *Thelaxes dryophila* (Schr.), apt. viv. – 2–3. *Mindarus obliquus* (Chol.), al. viv. (2) and apt. viv. (3). – 4. *Anoecia corni* (F.), al. viv. (fundatrigenia). – 5. *Eriosoma (Schizoneura) ulmi* (L.), al. viv. (fundatrigenia). – 6–7. *Tetraneura ulmi* (L.), fundatrix (6) and apt. virginogenia (7). – 8. *Pemphigus spyrothecae* Pass., ovip. female. – 9. *P. bursarius* (L.), apt. viv. (virginogenia). – 10. *Forda formicaria* v. Heyd., apt. viv. (10 ×).

Family Anoeciidae

Eyes of immature apterae with only three ommatidia; eyes of adult apterae often with more than three ommatidia; limit between head and pronotum distinct laterally behind the eyes. Antennae usually 6-segmented, in alatae with oval or subcircular secondary rhinaria. Processus terminalis shorter than 0.5 × base of ultimate segment. Marginal tubercles present on prothorax and some of the abdominal segments. Siphuncular pores placed on low cones. Cauda and anal plate rounded. Sexuales apterous, in dioecious species also dwarfish, buth with well developed rostrum.

Genus *Anoecia* Koch, 1857

Anoecia Koch, 1857: 275.
  Type-species: *Aphis corni* Fabricius, 1775.
Survey: 29.

Head and pronotum of apterae fused dorsally, but not laterally. Compound eyes of apterae usually well developed, with more than three ommatidia; immature apterae never with more than three ommatidia. Dorsum densely haired. Two types of dorsal hairs: pointed and spatulate (Fig. 111). Antennae usually 6-segmented, with oval or nearly circular secondary rhinaria on segm. III–VI in alatae, often also in apterae. Fore wing with very large, black pterostigma; media with one fork. Siphuncular pores rather wide; on low, haired cones; absent in first instar. Cauda and anal plate rounded. Prothorax and abd. segm. I–VII usually with marginal tubercles, but these may be absent from some segments on one or both sides. Male apterous. Oviparous female usually deposits 2 eggs.

A total of 21 species of *Anoecia* s. lat. are known. Some are dioecious with *Cornus* as the primary host, and grasses (Gramineae) as the secondary hosts. They feed on the leaves of the primary host and on the roots of the secondary hosts, and are visited by ants when living on the grass-roots. Other species are monoecious on Gramineae or Cyperaceae.

Key to species of *Anoecia*

Apterous viviparous females on primary host *(Cornus)*
(The spring generations of *major* and *nemoralis* are little known and not included).

1 Compound eyes large, consisting of many ommatidia. Antennae 6-segmented .... 9. *corni* (Fabricius) (fundatrigenia)
– Eyes small, triommatidia only. Antennae 5-segmented 2

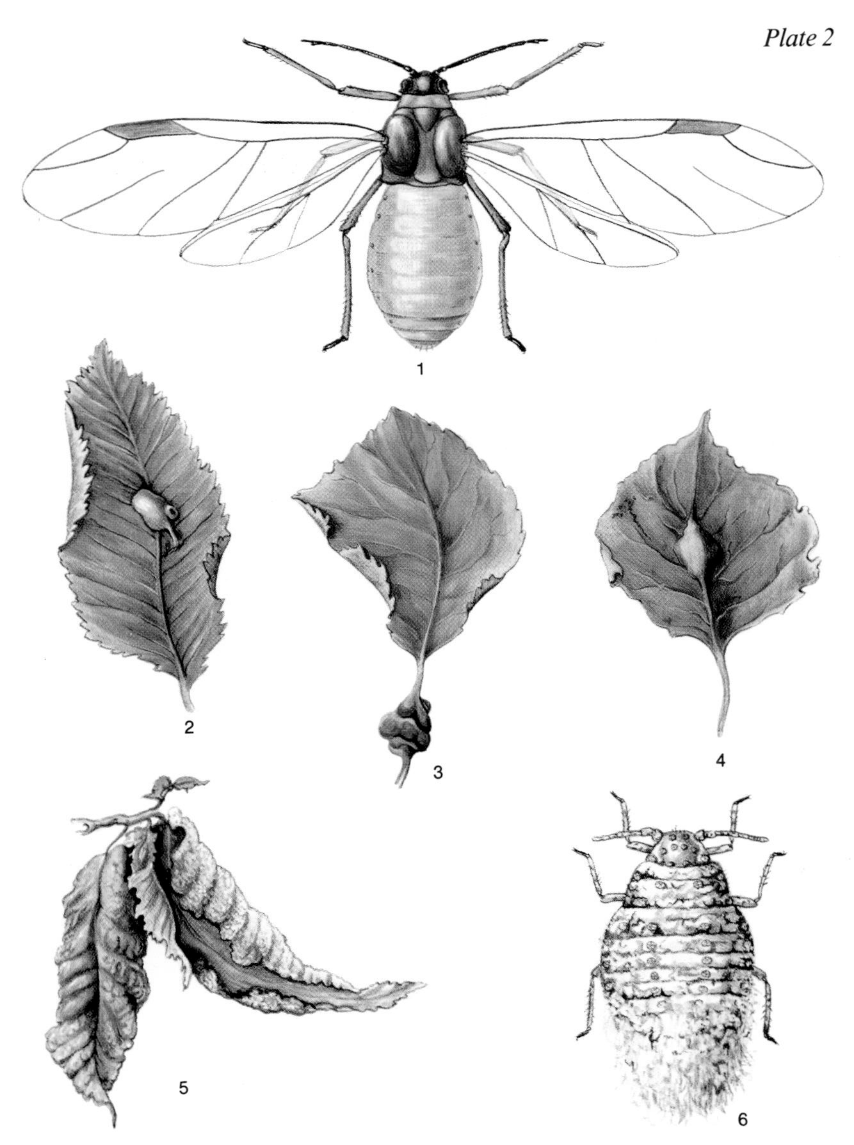

1. *Prociphilus (Stagona) xylostei* (DeGeer), al. viv. (fundatrigenia). – 2. *Tetraneura ulmi* (L.), gall on *Ulmus*. – 3. *Pemphigus spyrothecae* Pass., gall on *Populus*. – 4. *P. populinigrae* (Schr.), gall on *Populus*. – 5. *Eriosoma (Schizoneura) anncharlotteae* Dan., gall on *Ulmus carpinifolia*. – 6. *E. lanigerum* (Hausm.), apt. viv. (1 and 6: 10 ×) (5 drawn after photo in Danielsson; 6 partly after Bovien & Thomsen).

2 (1) All daughters are alate viviparous females without dorsal spot on abdomen. .... 12. *vagans* (Koch) (fundatrix)

– Offspring of the aptera may also be apterae. If nymphs have wing anlagen, then they will end up as adult alatae with large, black, dorsal spot on abdomen. .... 9. *corni* (Fabricius) (fundatrix)

Apterous viviparous females on herbaceous hosts

1 Processus terminalis about 0.11 × VIa (Fig. 121). Eye with only three ommatidia. .... 14. *pskovica* Mordvilko

– Processus terminalis longer than 0.2 × VIa. Eye often with more than three ommatidia. 2

2 (1) Abdomen not sclerotized. Eye usually with 2–10 (up to 25) ommatidia (Fig. 119). Secondary rhinaria absent. .... 13. *zirnitsi* Mordvilko

– Abdomen more or less sclerotized. Compound eyes large, usually with more than 25 ommatidia. Secondary rhinaria present or absent. 3

3 (2) Rostrum reaches to about mid abdomen. Spatulate hairs, if present, then on tergites VI and VII only. .... 10. *major* Börner

– Rostrum reaches to hind coxae only. Spatulate hairs may also be present on other tergites. 4

4 (3) Hairs on tergite III mainly spatulate, sometimes pointed, arranged in a very sparse, often incomplete transverse row. Marginal tubercles absent from abd. segm. V and VI .... 11. *nemoralis* Börner

– Hairs on tergite III spatulate or pointed, sometimes all pointed, not arranged in a transverse row. Abd. segm. I–VII usually all with marginal tubercles. 5

5 (4) Hairs on tergite IV arranged in a transverse row. Spatulate hairs present all over dorsum. Ant. segm. VI without secondary rhinaria. .... 12. *vagans* (Koch)

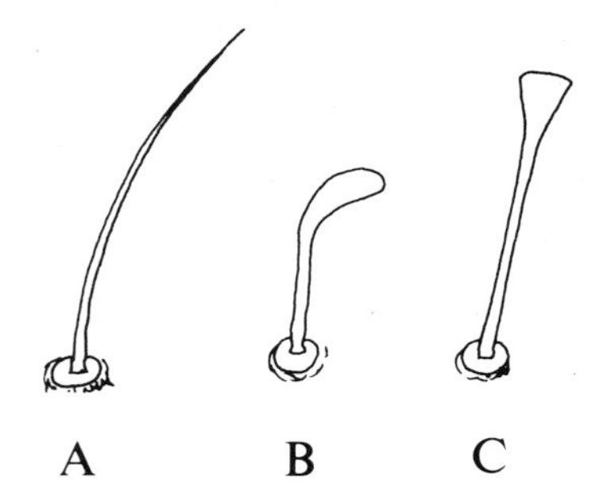

Fig. 111. Hairs of *Anoecia*. – A: pointed hair on abd. tergite VII of *A. vagans* (Koch) (al. viv. from October); B: spatulate hair on abd. tergite VII of same; C: spatulate hair on middle of abdominal dorsum of apterous virginogenia of *A. nemoralis* Börner.

– Hairs on tergite IV not arranged in a transverse row. Spatulate hairs present on posterior tergites, exceptionally also occurring in small numbers among the pointed hairs on the anterior tergites. Ant. segm. VI sometimes with a secondary rhinarium. .... 9. *corni* (Fabricius)

Alate viviparous females

(The alatae of *pskovica* and *zirnitsi* are unknown).

1 Abdomen without large, black, dorsal spot. .. 12. *vagans* (Koch) (born on *Cornus*)

– Abdomen with large, black, dorsal spot. 2

2 (1) Ant. segm. VI without secondary rhinaria (Fig. 113). Abd. tergite VII with spatulate hairs, sometimes only a single spatulate hair. .... 12. *vagans* (Koch) (born on secondary hosts)

– Ant. segm. VI often with secondary rhinaria. Abd. tergite VII without spatulate hairs. 3

3 (2) Marginal tubercles absent from abd. segm. V and VI. Spatulate hairs sometimes present on abd. segm. II and VI. Black dorsal spot on abdomen nearly bare. Ant. segm. III with 7–10 rhinaria. .... 11. *nemoralis* Börner

– Marginal tubercles usually present on abd. segm. V and VI. All body hairs pointed. Black dorsal spot on abdomen with hairs (0.02–0.03 mm long). 4

4 (3) Ant. segm. III with 9–17 rhinaria. Body 1.8–3.2 mm. .... 9. *corni* (Fabricius)

– Ant. segm. III with 13–22 rhinaria. Body 2.3–3.2 mm, usually about 3 mm. .... 10. *major* Börner

112

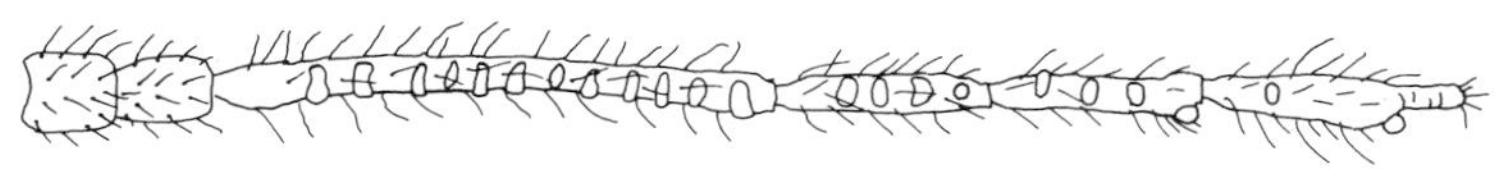

113

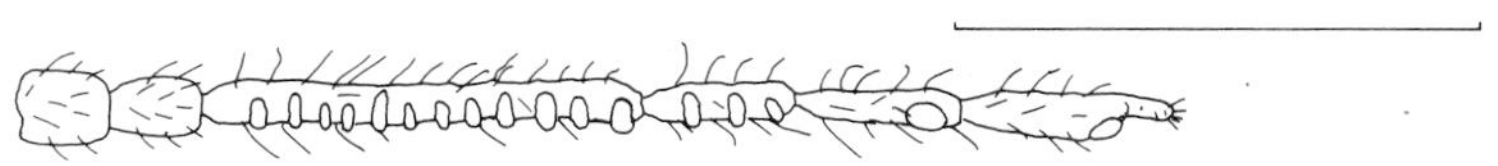

114

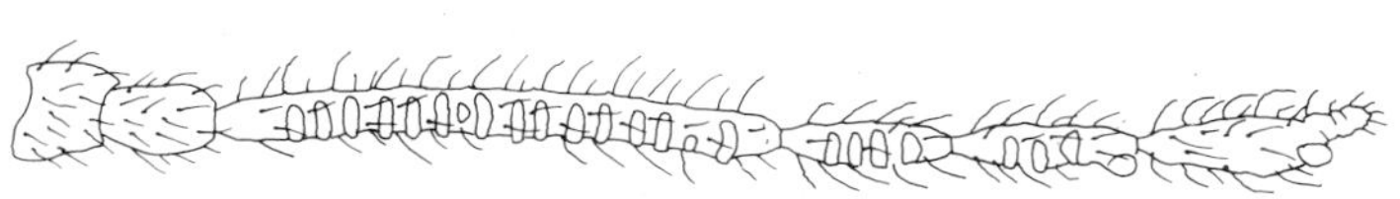

Figs. 112–114. Antennae of alate viviparous females of *Anoecia* spp. from autumn. – 112: *corni* (F.); 113: *vagans* (Koch); 114: *major* Börner. (Scale 0.5 mm).

Subgenus *Anoecia* Koch, 1857, s. str.

Processus terminalis longer than 0.2 × VIa.

9. ***Anoecia (Anoecia) corni*** (Fabricius, 1775)
Plate 1: 4. Figs. 112, 115, 116.

Aphis corni Fabricius, 1775: 214. – Survey: 30.

Fundatrix. Similar to the apterous fundatrigenia (see below), but differing in the following characters: eyes small, with only three facets (triommatidia); antennae 5-segmented, without secondary rhinaria.

Apterous viviparous female of 2nd or 3rd generation on primary host (apterous fundatrigenia). Dark brown. Dorsum sclerotized. Compound eyes large. Antennae 6-segmented, with secondary rhinaria. Round, flat marginal tubercles present on prothorax and abd. segm. I–VII, or on I–IV and VII. Arrangement and occurrence of pointed and spatulate hairs on dorsum variable.

Alate viviparous female from primary host (alate fundatrigenia). Head, thorax, antennae, legs, and cauda blackish brown. Abdomen bluish green with black dorsal spot

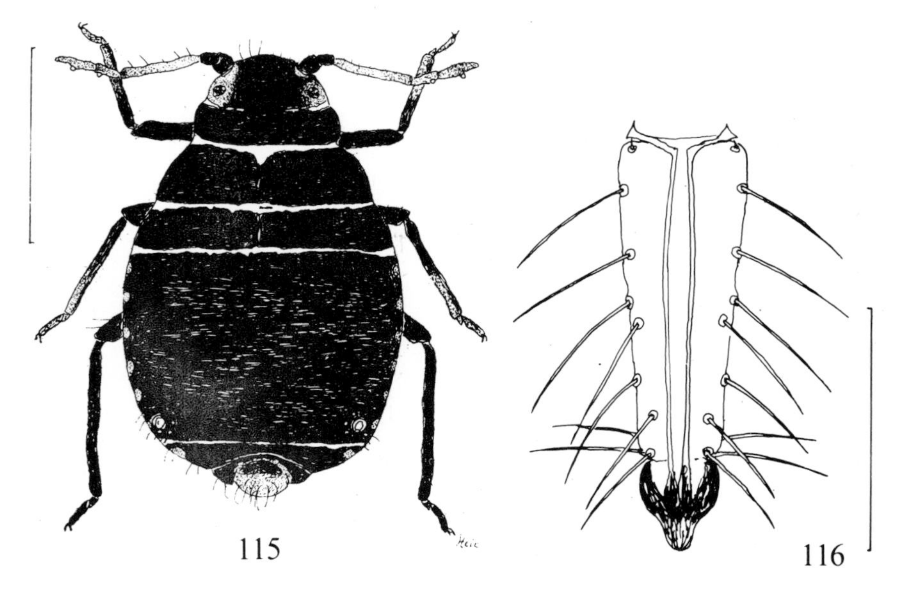

Fig. 115. Fundatrix of *Anoecia corni* (F.). (Scale 1 mm).

Fig. 116. Apical segment of rostrum of sexupara of same. (Scale 0.1 mm).

covering segm. III–VI and dark cross stripes on I–II and VII. Marginal tubercles as in apterous viviparous female. Dorsal hairs numerous, all pointed; hairs on middle of abdominal dorsum relatively short, 0.02–0.03 mm, and more dispersed than on the rest. Secondary rhinaria: ant. segm. III: 9–14, IV: 3–5, V: 2–4, VI: 0–2. Apical segm. of rostrum about 0.16–0.17 mm, about 0.89 × 2 sht (in sample from Finland). 2.4–2.7 mm.

Apterous viviparous female on secondary host (apterous virginogenia). Brown, sometimes greenish brown; sclerotized parts of abd. tergites III–VI often fused. Round, flat marginal tubercles usually present on prothorax and abd. segm. I–VII, sometimes absent from V and VI. Dorsum with numerous irregularly arranged, pointed hairs, 0.06–0.12 mm long; spatulate hairs present between the pointed hairs on tergites VI and VII, rarely also on V; hairs on (V–) VI–VII (–VIII) arranged in transverse rows. Eyes large, consisting of many ommatidia. Antennae 6-segmented, shorter than 0.5 × body; processus terminalis about 0.33 × VIa; ant. segm. III with 0–10 secondary rhinaria, IV with 1–4, V with 0–3, VI with 0–1. Rostrum reaching hind coxae. 1.9–2.8 mm.

Alate viviparous female from secondary host (alate virginopara and sexupara). Similar to alate fundatrigenia, but often with more secondary rhinaria: ant. segm. III: 10–17, IV: 3–5, V: 1–4, VI: 0–2. Apical segm. of rostrum 0.15–0.19 mm, 0.81–0.89 × 2 sht. Virginopara: 1.8–3.0 mm. Sexupara: 2.1–3.2 mm.

Oviparous female. Yellowish or brownish. Head, prothorax, and cauda dark. Body elongate. Marginal tubercles usually absent from abd. segm. V and VI, sometimes also from VII. Body hairs few, short. Eyes small, with only three facets (triommatidia). Antennae 5-segmented. 1.2–1.3 mm.

Male. Apterous. Slender. Dark brown. Marginal tubercles as in oviparous female. Body hairs numerous, pointed. Eyes large. Antennae 5-segmented. Siphuncular pores absent. About 0.8–0.9 mm.

Distribution. Common and widespread in Denmark; in Sweden occurring from Sk. north to Vb.; known from Norway; in Finland common in the South. – In nearly all parts of the northern hemisphere. In Europe from Scandinavia south to Spain, Portugal, and Yugoslavia, from Britain and France east to Russia; common in N Germany, Poland (incl. the Baltic region), and NW Russia. Asia: W Siberia, C Asia, Middle East, India, China, Korea, and Japan. Africa: N Africa; introduced in S Africa and N America.

Biology. The species has an obligatory host-alternation between *Cornus sanguinea,* the primary host, and various grasses (e. g. *Dactylis, Agrostis, Holcus, Calamagrostis, Brachypodium, Cynosurus),* the secondary hosts. Fundatrix and 2–3 fundatrigenia-generations develop on *Cornus* in spring. The alate fundatrigeniae bear young on grass-roots, and here several generations of apterous and alate viviparous females develop during the summer. In autumn all nymphs become alate sexuparae. They leave the grasses and fly to *Cornus* to bear the dwarfish sexuales. Anholocyclic overwintering on grass-roots does not take place in this species.

Note. The morphology and biology of *Anoecia corni* and other *Anoecia* spp. have been studied and described in detail by Zwölfer (1957). Measurements of the apical segm. of rostrum are made on rather few samples in my own collection. It is difficult to obtain exact information because several similar looking *Anoecia* spp. occur on the same hosts as *A. corni.* The diagnostic characters are also rather variable intraspecifically and only available in apterous viviparous females on secondary hosts. Isolated samples of aphids of the spring generations on *Cornus* cannot be identified. Some records of *corni* from Scandinavia and other regions may actually refer to *major, nemoralis,* or some other species.

10. ***Anoecia (Anoecia) major*** Börner, 1950
Figs. 62, 114.

Anoecia major Börner, 1950: 16. – Survey: 30.

Apterous viviparous female on primary host (fundatrix and apterous fundatrigenia). Similar to fundatrix and apterous fundatrigenia of *corni.* Diagnostic characters cannot be given.

Alate viviparous female from primary host (alate fundatrigenia). Unknown, probably very similar to *corni.*

Apterous viviparous female on secondary host (virginogenia). Very similar to *corni,* but rostrum longer, usually reaching abd. segm. II or nearly mid abdomen. Spatulate hairs rarely present on tergites VI and VII, always absent from V. On an average larger than *corni,* 2.2–3.0 mm.

Alate viviparous female from secondary host (sexupara). Very similar to *corni,* but larger, with numerous secondary rhinaria: ant. segm. III: 13–22, IV: 3–5, V: 1–3, VI: 0–2. Rostrum about 1.0 mm or longer (in *corni* shorter than 1.0 mm); apical segm. about 0.23 mm, 0.94–0.96 × 2 sht. 2.3–3.2 mm.

Distribution. Not in Denmark, Norway, or Finland; Sweden: Danielsson collected 1 al. viv. on *Salix caprea* (not a host plant) at Hällestad (Sk.). – Otherwise known from Germany and the Netherlands.

Biology. Very little is known about the biology of this species. In Germany it has been found on roots of *Phalaris (Calamagrostis) arundinacea* and *Brachypodium pinnatum.* Börner transferred it to *Cornus sanguinea* and acquired 2–3 generations of fundatrigeniae. The species has apparently not been found on *Cornus* in nature. Zwölfer (1957) found that *Agrostis* and *Dactylis,* the host plants of *corni,* were not accepted as hosts by apterous *major* found on *Phalaris.* The species is apparently not visited by ants.

Note. The description is based on Zwölfer (1957) and on 3 sexuparae from the Netherlands and one sexupara from Sweden. The length of the apical segm. of rostrum is based on these specimens only and not used in the key.

11. ***Anoecia (Anoecia) nemoralis*** Börner, 1950
Fig. 111 C.

Anoecia nemoralis Börner, 1950: 17. – Survey: 30.

Apterous viviparous female on primary host (fundatrix and apterous fundatrigenia). Similar to fundatrix and apterous fundatrigenia of *corni.* Diagnostic characters cannot be given.

Alate viviparous female from primary host (alate fundatrigenia). Unknown, probably very similar to *corni.*

Apterous viviparous female on secondary host (virginogenia). Green or blackish brown. Abd. segm. III–VI covered by dark sclerotized shield, other segm. with dark cross bands. Venter wax powdered. Marginal tubercles present on prothorax and abd. segm. I–IV and VII. Anterior part of dorsum with numerous irregularly arranged, pointed hairs, posterior part (from tergite III) with fewer hairs, arranged in transverse rows; spatulate hairs present on abd. segm. III–VII, sometimes also on metathorax and abd. segm. I–II, but then the spatulate hairs on III–VI are reduced. Compound eyes large. Antennae as in *corni;* segm. III with 0–2 secondary rhinaria, IV with 0–3, V with 0–3, VI with 0–2. Apical segm. of rostrum about 0.15 mm, about 0.85 × 2 sht. 1.6–2.3 mm.

Alate viviparous female from secondary host. Very similar to *corni,* but marginal tubercles always absent from abd. segm. V and VI, and the large, black, dorsal spot on abdomen is almost hairless. Dorsal hairs usually pointed, but spatulate hairs may occur on tergites II or VI. Secondary rhinaria: ant. segm. III: 7–10, IV: 2–4, V: 1–3, VI: 0–2. 2.2–2.5 mm.

Distribution. Not in Denmark, Norway, or Finland; recorded from the southern part of Sweden (Sk., Bl., Sm., Öl.). – Not in N Germany or the Baltic region of Poland, but recorded from Britain, France, Germany, Poland, Czechoslovakia, Hungary, and Russia (Ukraine).

Biology. The predominant kind of overwintering is the anholocyclic one as parthenogenetic females (adults and nymphs) on grass-roots in ants' nests. The species feeds on *Poa* spp., *Arrhenatherum elatius, Phleum pratense, Festuca* spp., *Agropyrum repens, Alopecurus geniculatus, Bromus erectus, Deschampsia* spp., *Sieglingia decumbens,* and *Panicum antidotale* (Zwölfer 1957, Stroyan 1964), and the colonies are visited and cared for by ants, most frequently *Lasius flavus* and *L. niger.*

Cornus sanguinea is the primary host, but most alate viviparous females produced by grass-feeders in autumn fly to other grass plants. Several colonies do not at all produce alatae in autumn. Zwölfer found that only 3.5% of the alate sexuparous *Anoecia* arriving at *Cornus* bushes in autumn showed the morphological characters of *nemoralis,* although 25% of the *Anoecia* populations found on secondary hosts in the area (in Germany) belonged to this species. The fundatrix and the fundatrigeniae on *Cornus* have not been described (Zwölfer 1957: 216: "Über die Hauptwirtgenerationen dieser vorwiegend anholozyklischen Art liegen keine Beobachtungen vor"). The possibility thus exists that some records of *corni* from *Cornus* refer to *nemoralis,* but it seems insignificant because Zwölfer could not observe any increase of population density on grasses in early summer, when arrivals of alate migrants from *Cornus* might be expected.

12. ***Anoecia (Anoecia) vagans*** (Koch, 1856)
Figs. 111 A & B, 113.

Schizoneura vagans Koch, 1856: 268. – Survey: 30.

Fundatrix. Morphological characters as in fundatrix of *corni*, but all daughters become alate viviparous females.

Alate viviparous female from primary host (fundatrigenia). Differs from *corni* by the absence of the black, dorsal spot on abd. segm. III–VI. Each of the tergites I–VII with a transverse row of short (0.007–0.015 mm), often spatulate hairs. Secondary rhinaria: ant. segm. III: 6–10, IV: 1–3, V: 0–2, VI: 0. 2.3–2.8 mm.

Apterous viviparous female on secondary host. Yellowish. Marginal tubercles as in *corni*. Hairs on tergites IV–VIII arranged in transverse rows; spatulate hairs present all over the dorsum, but reduced on tergites IV and V; pointed hairs occur on anterior part of body (from tergite III), on head being much more frequent than spatulate hairs. Siphuncular cones with strong, curved, spatulate hairs. Antennae 6-segmented, shorter than 0.5 × body; secondary rhinaria: ant segm. III: 0–5, IV: 1–3, V: 0–1, VI: 0. 2.1–2.7 mm.

Alate viviparous female from secondary host (virginopara and sexupara). Contrary to fundatrigenia with black, dorsal spot on abdomen. Spatulate hairs present only on tergite VII. The same number of secondary rhinaria in virginopara as in fundatrigenia; the sexupara with more of them: ant. segm. III: 10–14, IV: 2–5, V: 0–2, VI: 0. Apical segm. of rostrum about 0.17 mm, 0.7–0.8 × 2 sht. Virginopara: 2.1–2.5 mm. Sexupara: 2.5–3.9 mm.

Distribution. In Denmark recorded from NEZ, but probably more widespread and not uncommon; in Sweden known from several districts, from Sk. in the South to P. Lpm. in the North, including Öl.; not recorded from Norway or Finland. – Widespread in Europe through C Europe to Portugal and Bulgaria, and from Britain and N Germany to Poland and NW Russia. Also N Africa.

Biology. The species is holocyclic with obligatory host-alternation between *Cornus sanguinea* and grass-roots, as in *corni*, but only one fundatrigenia-generation is born on *Cornus*. Among the secondary hosts are cereals as barley (*Hordeum sativum*), wheat (*Triticum sativum*) and wild grasses as *Agropyrum repens* and *Arrhenatherum elatius*. The grass-feeders are sometimes visited by ants.

13. ***Anoecia (Anoecia) zirnitsi*** Mordvilko, 1931
Figs. 117–119.

Anoecia zirnitsi Mordvilko, 1931: 1314. – Survey: 31.

Apterous viviparous female. Greyish green or brownish green. Abdominal dorsum membranous with small intersegmental muscle sclerites, sometimes with more or less fused, weakly sclerotized, irregularly arranged sclerites. Marginal tubercles present on prothorax and abd. segm. I–VII, now and then also on meso- and metathorax; may be indistinct on abd. segm. V and VI; those on prothorax very large (diameter 0.08–0.10); those on abdomen formed by polygonal facets. Dorsal hairs pointed, on tergite III up to 0.11 mm (5.7 × IIIbd.), irregularly arranged on most tergites, in transverse rows on tergites VII and VIII; spatulate hairs and forked hairs sometimes present over most of dorsum. Eyes rather small, sometimes only triommatidia present, in other individuals also compound eyes consisting of 2–10(–25) ommatidia. Antennae 5- or 6-segmented; segm. III in 6-segmented antenna about as long as VI.; secondary rhinaria absent. Rostrum reaching hind coxae. 1.5–2.1 mm.

Oviparous female. With a pair of large, roundish, facetted wax gland fields on sternites IV–VI. Otherwise similar to the apterous viviparous female.

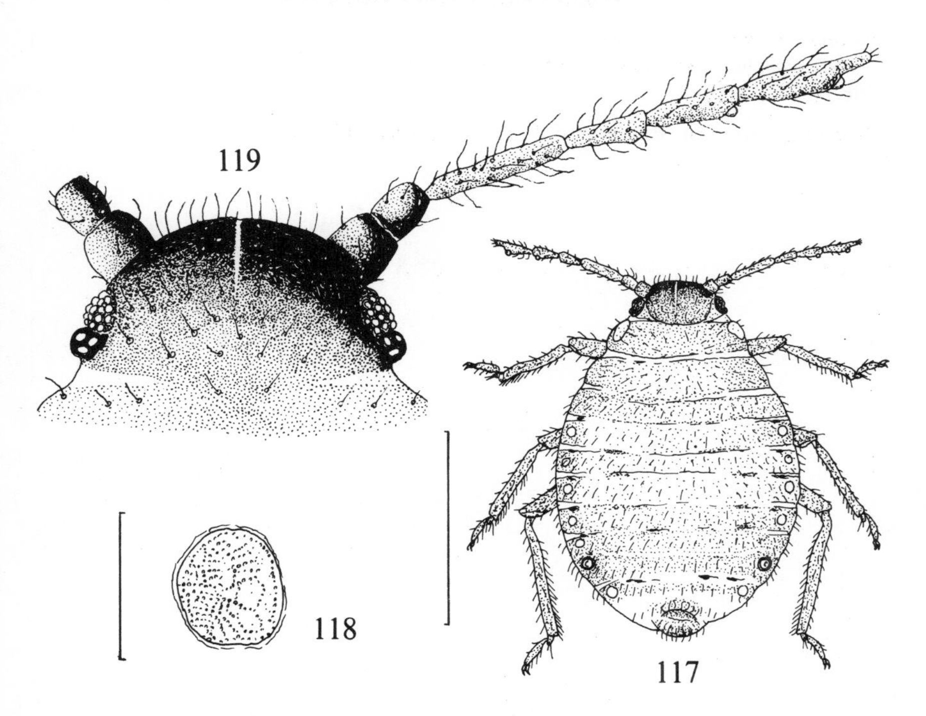

Figs. 117–119. *Anoecia zirnitsi* Mordv., apt. viv. – 117: body; 118: marginal tubercle of abd. segm. IV; 119: head and one antenna. (Scales 1 mm for 117, 0.05 mm for 118).

Distribution. In Denmark found on Samsø (EJ, Gissel Nielsen coll.); in Sweden known from Sk., Sm., and Öl.; not in Norway or Finland. – Recorded from Britain, Germany, Poland, and the European USSR (Latvia, Russia). Unknown in N Germany and the Baltic region of Poland.

Biology. The species is holocyclic and monoecious. Colonies are found on thin roots of *Festuca, Agrostis, Brachypodium, Lolium,* and other grasses. The oviparous females deposit the eggs in October–December. The eggs are collected and stored by ants *(Lasius flavus)* in "packets" containing up to 900 eggs (Zwölfer 1957). A few viviparous females can still be observed in December (in Germany), but the overwintering takes place only as eggs. In spring the aphids occur in the ants' nests, later at some distance from the nests. Alate viviparous females and males are unknown. Zwölfer suggests that the oviparous female may be parthenogenetic, which is very exceptional in aphids.

Subgenus *Paranoecia* Zwölfer, 1957

Paranoecia Zwölfer, 1957: 198.
Type-species: *Anoecia pskovica* Mordvilko, 1916.
Survey: 335.

Processus terminalis very short, only about 0.1 × VIa (Fig. 121).

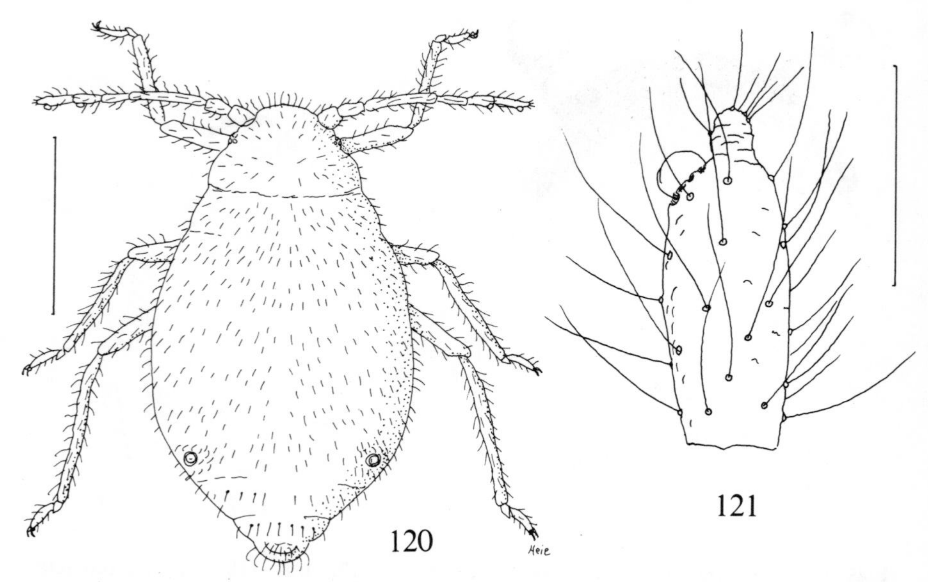

Figs. 120, 121. *Anoecia (Paranoecia) pskovica* Mordv. – 120: apterous viviparous female; 121: ant. segm. VI of same. (Scales 1 mm for 120, 0.1 mm for 121).

14. ***Anoecia (Paranoecia) pskovica* Mordvilko, 1916**
Figs. 120, 121.

Anoecia pskovica Mordvilko, 1916: 76. – Survey: 31.

Apterous viviparous female. Greyish white. Abdomen without dorsal sclerotizations. Marginal tubercles present on prothorax and abd. segm. I–VII. Dorsal hairs long (0.05–0.13 mm) and pointed. Eyes small, triommatidia only. Antennae 6-segmented, about 0.5 × body; processus terminalis about 0.11 × VIa; secondary rhinaria: III: 0–6, IV: 0–3, V: 0–2, VI: 0. Rostrum reaches to abd. segm. III or IV. 2.4–2.7 mm.

Oviparous female. With one pair of large, roundish or oval, facetted wax gland fields on sternites IV–VIII. Very similar to the apterous viviparous female, also with regard to size, but antennae and rostrum relatively shorter.

Male. Apterous. Slender. Greyish white. Antennae 5-segmented, without secondary rhinaria. Siphuncular pores absent. Otherwise similar to the females. 1.5–2.6 mm.

Distribution. In Denmark found at Femmøller and on Samsø (EJ); in Sweden known from Sk. and Sm.; not in Norway or Finland. – Otherwise found in Britain, Germany (but not in N Germany), and NW Russia.

Biology. The species is holocyclic and monoecious. It lives on roots and subterranean runners of *Carex* and *Eriophorum*, mostly in humid soil; in Denmark also on *Carex arenaria* in rather dry soil. Alate viviparous females are very scarce, apparently only observed in Russia by Mordvilko. Also males are comparatively rare. The oviparous females oviposit in November–December in S Germany. The eggs, which are covered with wax from the ventral glands of the oviparous female, are not collected and stored by ants, but the species is usually attended by ants and can often be found in nests of *Lasius flavus*, kept in special "cow-houses". Overwintering of viviparous females is of minor importance (Zwölfer 1957).

Family Pemphigidae

Eyes of apterous individuals and young nymphs usually with only three ommatidia (except in apterous adults of some species of Fordinae and alatiform apterae of Eriosomatinae). Limit between head and pronotum usually distinct. Antennae short, usually 6-segmented, with very short processus terminalis. Secondary rhinaria transverse oval, subcircular, or narrow and ringlike; usually absent from apterous specimens. Media of fore wing with one fork or unbranched; hind wing with one or two oblique veins. Siphunculi pore-shaped or absent. Cauda broad, rounded, not constricted. Wax glands often well developed, usually occurring in groups on several roundish, facetted plates.

The sexuales are apterous and dwarfish as in Mindaridae and Anoeciidae, but they have no rostrum; this is the chief distinguishing character of Pemphigidae. The sexuparae contain embryos without stylets. The stylets of embryos are usually easy to observe in mounted viviparous females, so their absence from an alata full of embryos shows that it is a sexupara. The oviparous female lays only one egg, which is almost as big as the female itself (Fig. 42).

Most species are dioecious, producing galls or leaf-curling on the primary hosts. Some species migrate to roots of woody secondary hosts, others to roots or aerial parts of herbaceous secondary hosts. Many can hibernate as parthenogenetic females and may do so winter by winter. In this way the species can survive in areas outside the geographical range of the primary host. Some of these species have holocyclic populations in warm climates, e.g. in the Mediterranean region, and anholocyclic populations in cold climates, e.g. in Scandinavia. Others are exclusively anholocyclic on plants supposed to be the secondary hosts of remote ancestors. A few species are monoecious and holocyclic on trees identical with or related to trees used as primary hosts by related species.

The taxonomy is difficult because the various generations of the same species usually are very different. It may, on the other hand, be difficult or even impossible to separate species on the basis of specimens of the same morph. In some groups a reliable identification depends on studies of life cycles and host plant affinities.

Keys to subfamilies of Pemphigidae

Apterous viviparous females on secondary hosts

1 Wax gland plates with one or more central facets surrounded by a ring of rather uniform facets (Figs 128 A–E, 147). Siphuncular pores usually present. .... **Eriosomatinae** (p. 114)
– Wax gland plates with several rather uniform facets not arranged in a ring surrounding a central area (Figs 198, 207), or absent. Siphuncular pores absent. 2
2 (1) Tarsi one-segmented. 3
– Tarsi two-segmented. 4
3 (2) Processus terminalis about 0.3 × basal part of ultimate antennal segment. .... **Fordinae: Melaphidini** (p. 206)
– Processus terminalis much shorter, reduced to a vanishing point. .... **Pemphiginae** (partim: *Mimeuria* p. 163)
4 (3) First segm. of hind tarsus with two hairs. Primary rhinaria surrounded by short hairs. Dorsum with 2–4 longitudinal rows of wax gland plates on abd. segm. (III–)IV–VII. .... **Pemphiginae** (p. 139)
– First segm. of hind tarsus with four or more hairs. Primary rhinaria not surrounded by short hairs, or if such hairs are present, then dorsum with six longitudinal rows of wax gland plates, or wax gland plates absent. .... **Fordinae: Fordini** (p. 192)

Apterous viviparous females (including fundatrices) **on woody plants** (in Scandinavia) (key based on host plants).

1 On *Ulmus*. 2
– Not on *Ulmus*. 3
2 (1) In galls on leaves. .... **Eriosomatinae** (on primary hosts) (p. 114)
– On roots. .... **Pemphiginae** (partim: *Mimeuria ulmiphila* on secondary host) (p. 163)
3 (2) On Pomaceae. 4
– Not on Pomaceae. 5
4 (3) On leaves. .... **Pemphiginae** (*Prociphilus pini* on primary host) (p. 159)
– On roots, branches, or trunks. .... **Eriosomatinae** (on secondary hosts) (p. 114)
5 (3) On aerial parts (leaves, branches, shoots) of *Populus, Fraxinus, Lonicera*, or *Acer*. .... **Pemphiginae** (on primary hosts) (p. 139)
– On roots of various other trees or bushes. 6
6 (5) On roots of *Ribes*. .... **Eriosomatinae** (on secondary hosts) (p. 114)
– On roots of coniferous trees or *Salix*. ... **Pemphiginae** (on secondary hosts (p. 139)

Alate viviparous females

1 Secondary rhinaria circular, subcircular, or transverse oval. Media of fore wing unbranched. .... **Fordinae: Fordini** (p. 192)
– Secondary rhinaria narrow, stripe-shaped or ringlike. Media of fore wing unbranched or with one fork. 2
2 (1) Pterostigma rather long drawn out, its posterior margin concave (Fig. 245). Ultimate ant. segm. with ringlike secondary rhinaria (Fig. 247). .... **Fordinae: Melaphidini** (p. 206)
– Pterostigma not long drawn out; its posterior margin not conspicuously concave, or, if pterostigma of such shape, then ultimate ant. segm. without secondary rhinaria (Fig. 179). 3
3 (2) Hind wing with two oblique veins with bases very close to each other (Figs. 182, 189). Wax gland plates honeycomb-like, without a ring of rather uniform facets surrounding a central area (Figs 162, 198). With three rudimentary gonapophyses. Primary host not *Ulmus*. .... **Pemphiginae** (p. 139)
– Hind wing with two oblique veins well separated at bases (Fig. 134), or with only one oblique vein (Fig. 136). Wax gland plates with a ring of rather uniform facets surrounding a central area with one or more facets (Figs. 127 A–E, 147). With two rudimentary gonapophyses, or the gonochaetes not placed on distinct gonapophyses. Primary host *Ulmus*. ...... **Eriosomatinae** (p. 114)

SUBFAMILY ERIOSOMATINAE

Alate females with narrow, ringlike secondary rhinaria. Media of fore wing with one fork or unbranched. Hind wing with 1 or 2 oblique veins. First tarsal segm. with 3-3-3 or 3-2-2 hairs in alatae and 2-2-2 in apterae, or tarsi one-segmented. Siphuncular pores present or absent, usually present in adults living on secondary hosts. Wax gland plates in apterous virginogeniae with one or more central facets surrounded by a ring of facets (Fig. 127 A–E).

Most species are holocyclic and dioecious and use *Ulmus* as the primary host. The secondary hosts are various trees, bushes, or herbs.

The family-group names Schizoneurinae and Schizoneurini ("Schizoneuriden" of Herrich-Schaeffer in C. L. Koch, 1857; type-genus *Schizoneura* Hartig, 1839) are valid names for subfamily and tribe, if the Code shall be followed strictly, because *Schizoneura*, according to Eastop & Hille Ris Lambers (1976), is not a synonym of *Eriosoma* but a subgenus of that genus. According to Art. 42 (a) subgenus-names belong to the category of genus-names and are thus available as type-genera for family-groups.

The family-group names with *Eriosoma* Leach, 1818 as type-genus are younger and have been in common use since 1920, when Baker (who regarded *Schizoneura* as a synonym of *Eriosoma*) introduced Eriosomatinae. They have been retained in this work as a case which is exemplified in Art. 40(b), example 2 of the Code. Of the same reason usage of the older name Tetraneurinae ("Tetraneuriden" Herrich-Schaeffer in Koch, 1857) should be avoided, even if it may become practice to synonymize *Schizoneura* with *Eriosoma* in future.

Key to tribes of Eriosomatinae

Spring and summer generations on *Ulmus*

1 In curled or rolled leaves or in large, blister-shaped galls placed directly on thin branches. .... **Eriosomatini** (p. 115)

– In more or less closed galls placed on the surface of the leaves. **Tetraneurini** (p. 130)

Alate viviparous females

1 Media of fore wing with one fork. Hind wing with two oblique veins. Ant. segm. III distinctly longer than segm. IV+V+VI. Siphuncular pores present. .... **Eriosomatini** (p. 115)

– Media of fore wing with one fork or unbranched. Hind wing with one or two oblique veins. Ant. segm. III as long as or shorter than segm. IV+V+VI. Siphuncular pores present or absent. .... **Tetraneurini** (p. 130)

Apterous viviparous females not found on *Ulmus*.

1 Four longitudinal rows of wax gland plates on dorsum of body (Fig. 123). Tarsi 2-segmented. On Pomaceae, Ribesiaceae, or Compositae. .... **Eriosomatini** (p. 115)

– Four to six longitudinal rows of wax gland plates on dorsum of body. Tarsi 1-segmented, or – if 2-segmented – then 6 wax gland plates present on most body segments (Fig. 142). On Gramineae, Cyperaceae, or Labiatae. .... **Tetraneurini** (p. 130)

TRIBE ERIOSOMATINI

Media of fore wing with one fork. Hind wing with two oblique veins. Ant. segm. III of alatae distinctly longer than segm. IV+V+VI. Siphuncular pores absent in fundatrices, but present in other viviparous morphs. Apterae on secondary hosts with 4 longitudinal rows of wax gland plates. Adults with two tarsal segments.

The dioecious species curl the leaves or produce large, blister-shaped galls on *Ulmus*, the primary host.

Genus *Eriosoma* Leach, 1818

Eriosoma Leach, 1818: 60.
Type-species: *Eriosoma mali* Leach, 1818.
= *Aphis lanigera* Hausmann, 1802.
Survey: 190.

Six species in Scandinavia, 20 species in the world. Danielsson (1979) has given thorough descriptions of all Scandinavian species except *E. (Schizoneura) lanuginosum* (Hartig).

Key to subgenera of *Eriosoma*

1 First instar nymphs with smooth or slightly imbricate tarsi and long dorsoapical hairs on second tarsal segments. ...... *Eriosoma* s. str. (p. 118)

– First instar nymphs with spinulose tarsi and normal dorsoapical hairs on second tarsal segments. .... *Schizoneura* Hartig (p. 121)

Key to species of *Eriosoma* s. lat.
(partly after Danielsson (1979))

Apterous viviparous females
(except virginogeniae of *E. (S.) lanuginosum* (Hartig)).

1 On *Ulmus.* 2
– On other plants. 5
2 (1) In large, closed, blister-shaped leaf-galls placed directly on the branches (Fig. 128). Fundatrix bluish black, with 5- segmented antennae. Fundatrigeniae olive green, with 6-segmented antennae. .... 17. *lanuginosum* (Hartig)
– Not in closed galls, but in curled or rolled leaves. All adult individuals with 6-segmented antennae, or – if 5-segmented – then segm. III usually with an incipient subdivision. 3
3 (2) Infested leaves curled downward on one side forming yellowish leaf-rolls (Fig. 135), not clustered together on certain shoots. Fundatrix dark green, almost black. .... 20. *ulmi* (Linné)
– Infested leaves curled downwards on both sides or completely distorted, green, yellowish or reddish, often clustered together on certain shoots. Fundatrix green to dark bluish grey. 4
4 (3) Apical segm. of rostrum longer than 0.20 mm. ..... 18. *patchiae* (Börner & Blunck)
– Apical segm. of rostrum shorter than 0.20 mm. ...... 16. *anncharlotteae* Danielsson
5 (1) Apical segm. of rostrum longer than 0.20 mm, with 15 or more accessory hairs. On *Senecio* or *Cineraria.* .... 18. *patchiae* (Börner & Blunck)
– Apical segm. of rostrum shorter than 0.20 mm, with less than 15 accessory hairs. Not on *Senecio* or *Cineraria.* 6
6 (5) Wax gland plates with 0–7 central cells, much smaller than surrounding cells (Fig. 127 A). Empodial hairs longer than claws. .... 15. *lanigerum* (Hausmann)
– Wax gland plates with a central cell much larger than surrounding cells (Fig. 127 B–E). Empodial hairs not reaching tips of claws. 7
7 (6) Cauda with 3–5 hairs. Antenna 5-segmented, shorter than 0.19 mm. Empodial hairs very short, hardly visible. ..... 19. *sorbiradicis* Danielsson
– Cauda with 2 hairs. Antenna 6-segmented, longer than 0.19 mm. Empodial hairs distinctly visible. 8
8 (7) Central cell of wax gland plates defined by a distinct double line (Fig. 127 C). Ant. segm. VI 0.45–0.58 × apical segm. of rostrum. Longest apical hair on processus terminalis 0.015–0.019 mm long. Empodial hairs shorter than half length of claws. .... 20. *ulmi* (Linné)
– Central cell of wax gland plates defined by a greyish band, sometimes with a narrow dark line, but not with a double line (Fig. 127 B). Ant. segm. VI 0.31–0.45 × apical segm. of rostrum. Longest apical hair on processus terminalis shorter than 0.015 mm. Empodial hairs about half as long as claws or longer. .... 16. *anncharlotteae* Danielsson

Alate viviparous females
(except *E. (S.) sorbiradicis* Danielsson)

1 Penultimate ant. segm. without secondary rhinaria. 2
– Penultimate ant. segm. with secondary rhinaria. 5
2 (1) Fundatrigeniae (spring and summer). 3
– Sexuparae (autumn). 4
3 (2) Antenna 1.00 mm or shorter; segm. III shorter than 0.60 mm, with 27–33 secondary rhinaria; segm. VI shorter than 3.0 × maximal width of the segment (Fig. 122 D). .. 16. *anncharlotteae* Danielsson

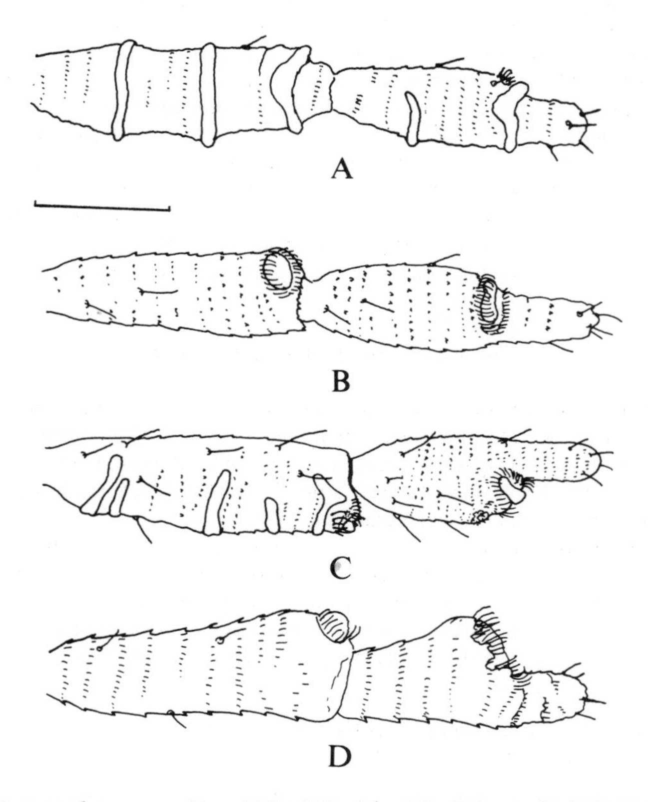

Fig. 122. Antennal segments V and VI of al. viv. of A: *Erisoma lanigerum* (Hausm.); B: *E. (Schizoneura) ulmi* (L.); C: *E. (S.) patchiae* (Börn. & Bl.); D: *E. (S.) anncharlotteae* Danielsson. – A is an alata from *Malus* collected in autumn, B–D are alate fundatrigeniae from *Ulmus*. (B and C after Marchal, redrawn). (Scale 0.1 mm).

– Antenna longer than 1.10 mm; segm. III longer than 0.60 mm, with 30–46 secondary rhinaria; segm. VI longer than 3.0 × maximal width of the segment (Fig. 122 B). .... 20. *ulmi* (Linné)

4 (2) Ant. segm. III with 22–33 secondary rhinaria. Longest hair on ant. segm. III as long as IIIbd. or a little longer. Abdomen with small marginal wax gland plates on segm. I–VII, sometimes also a spinal pair on tergite VIII. .... 16. *anncharlotteae* Danielsson

– Ant. segm. III usually with less than 22 secondary rhinaria, but in some specimens with up to 27. Longest hair on ant. segm. III about 3 × IIIbd. Abdomen with distinct marginal wax gland plates, consisting of a transparent central cell surrounded by a sclerotized area. Tergites I–VII also with a pair of spinal wax gland plates, but these often reduced to dark spots without the central area. .... 20. *ulmi* (Linné)

5 (1) Ant. segm. V as long as IV or a little shorter. Antennae always 6-segmented. .... 17. *lanuginosum* (Hartig)

– Ant. segm. V longer than IV, or antennae not 6-segmented. 6

6 (5) Primary rhinarium on penultimate ant. segm. not surrounded by short hairs (Fig. 122 A). .... 15. *lanigerum* (Hausmann)

– Primary rhinarium on penultimate antennal segment surrounded by short hairs (Fig. 122 C). .... 18. *patchiae* (Börner & Blunck)

Subgenus *Eriosoma* Leach, 1818, s. str.

Primary rhinaria of fundatrices (of holocyclic species, which do not occur in Scandinavia) not surrounded by short hairs; such hairs present in virginogeniae and around the accessory rhinaria close to the primary rhinarium on the distal ant. segm. of alatae (Fig. 122 A). First instar nymphs with smooth or slightly imbricate tarsi and long dorsoapical hairs on second tarsal segm.

The subgenus consists of six or seven species, but a revision is needed because several taxonomic problems are unsolved. Most species are nearctic. In America several holocyclic and dioecious species occur, with *Ulmus* as the primary host, and members of Pomaceae as the secondary hosts. One cosmopolitan, anholocyclic species, which is a severe pest to apple trees, is known from Scandinavia.

15. ***Eriosoma (Eriosoma) lanigerum*** (Hausmann, 1802)

Plate 2: 6. Figs. 122 A, 123, 124, 127 A & I.

Aphis lanigera Hausmann, 1802: 440. – Survey: 191.

Apterous viviparous female. Purple, red or brown; covered with white wool produced by distinct wax gland plates, each consisting of 0–7 small central facets surrounded by a ring of larger facets; several such plates present on head, one spinal pair and one marginal pair present on each of the thoracic and abdominal segm.; tergite VIII only with a spinal pair. Antennae usually 6-segmented, 0,17–0.24 × body; segm. III usually shorter than IV+V+VI, but longer than IV+V, sometimes fused with IV, so that the antenna becomes 5-segmented; VI about as long as V and longer than IV, 0.42–0.52 × apical segm. of rostrum. Longest hair on ant. segm. III shorter than IIIbd. Apical hairs on processus terminalis 0.008–0.011 mm. Rostrum reaches past middle coxae; apical segm. 1.36–1.83 × 2sht., with 7–9 accessory hairs. Siphuncular pores present on low warts. Cauda with 2 hairs. 1.6–2.4 mm.

Alate viviparous female. Abdomen reddish brown. Wax gland plates very small and indistinct, usually one spinal and one marginal pair on each of the abd. segm. Antennae usually 6-segmented, but segm. III+IV may be fused; sometimes even all segments of flagellum are fused; antenna about 0.4 × body; segm. III about 1.25 × the length of IV+V+VI; IV shorter than V, usually a little longer than VI; processus terminalis about 0.3 × VIa. Secondary rhinaria: III: 16–26, IV: 3–6, V: 3–7, VI: 0–3. Apical segm. of rostrum about 1.2–1.3 × 2sht. Siphuncular pores present on dark, low, hairy warts. 1.6–2.3 mm.

Sexuales. Antennae 5-segmented. Dwarfish, the oviparous female about 1 mm, the male about 0.5 mm.

Distribution. Rather common and widespread in Denmark, but control measures usually keep the populations on a low level so that the species is common only locally in some years; in Sweden recorded from the southern part, north to Dlsl. and Ög., but due to regulations about its control since 1945 now found regularly only in western Sk.;

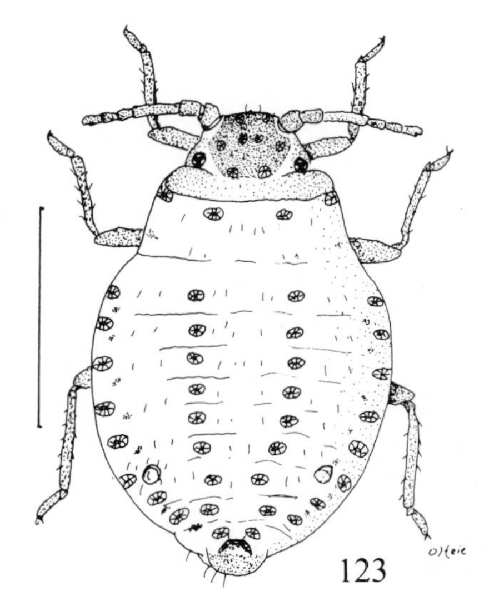

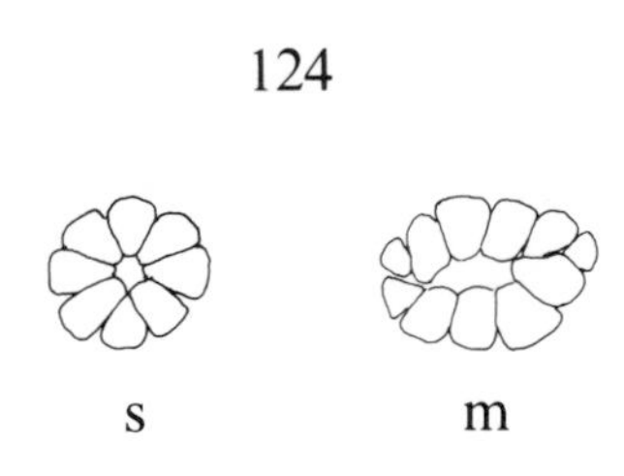

Figs. 123, 124. *Eriosoma lanigerum* (Hausm.). – 123: apt. viv.; 124: wax gland plates of same; s = spinal plate on abd. segm. III; m = marginal plate on abd. segm. IV. (Scale 1 mm for 123).

in Norway known from Ry and MRy; not known from Finland. – Cosmopolitan; in Europe occurring from the British Isles to the Baltic region of Russia, not rare in N Germany, common in Poland, also at the Baltic coast, in Russia north to the January isotherm of ÷3–÷4°C., south to the Mediterranean Sea; in Asia recorded from the Middle East, the Himalayas, C Asia, and China; in Africa recorded from Egypt, Kenya, Rhodesia, and S Africa; also known from SE Australia, Tasmania, New Zealand, and Hawaii; widespread in the USA and Canada, recorded from Mexico, Haiti, and several countries in S America. The origin is unknown and must be searched for outside Europe, where it was first found in England in 1787; later it appeared in Germany (1802), France (1812), and other European countries. It seems to have invaded Denmark as late as in the beginning of the 20th century according to Boas (1923), and arrived at Sweden in 1930. It has been suggested that the species originates from N America and might be identical with a holocyclic species with *Ulmus americana* as the primary host and apple as the secondary host, but no convincing proof has been published. The species may be introduced in America as well.

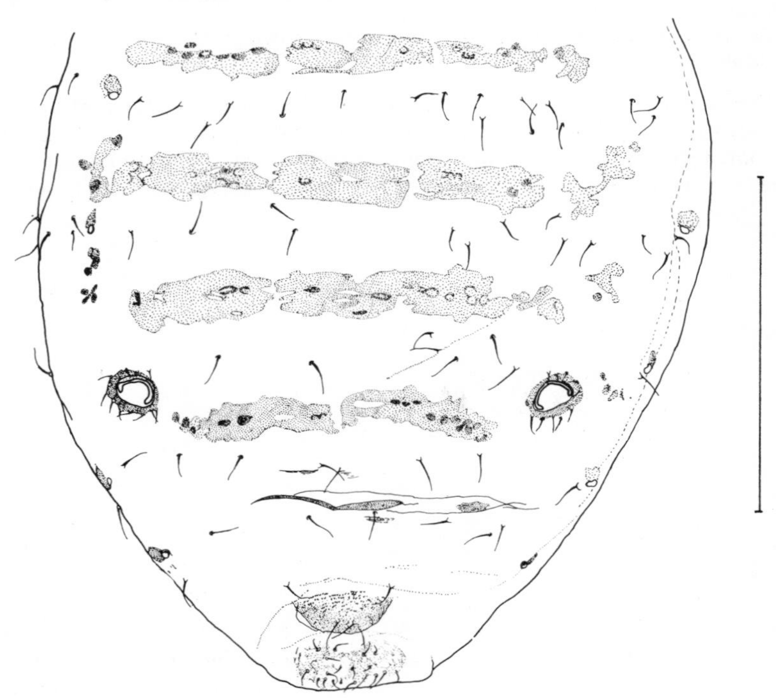

Fig. 125. *Eriosoma (Schizoneura) anncharlotteae* Dan., abdomen of alate fundatrigenia. (After Danielsson). (Scale 0.5 mm).

Biology. The species lives on roots, trunks, and branches of apple *(Malus)* and is a severe pest. In Sweden it has also been found on pear *(Pyrus communis)* and *Cotoneaster horizontalis* (Danielsson 1979). The branches are deformed and may show cancer-like swellings. Overwintering of apterous viviparous females and young takes place on *Malus*. The species seems to be exclusively anholocyclic though alate sexuparae able to bear sexuales are born in autumn. I have not observed them myself. Attempts to transfer them to *Ulmus* have failed according to various authors; if the eggs laid by the oviparous females hatch at all, the young fundatrices will not reach maturity.

Subgenus *Schizoneura* Hartig, 1839

Schizoneura Hartig, 1839: 654.
Type-species: *Chermes ulmi* Linné, 1758.
Survey: 390.

Primary rhinaria surrounded by short hairs in all morphs. Wax gland plates more or less

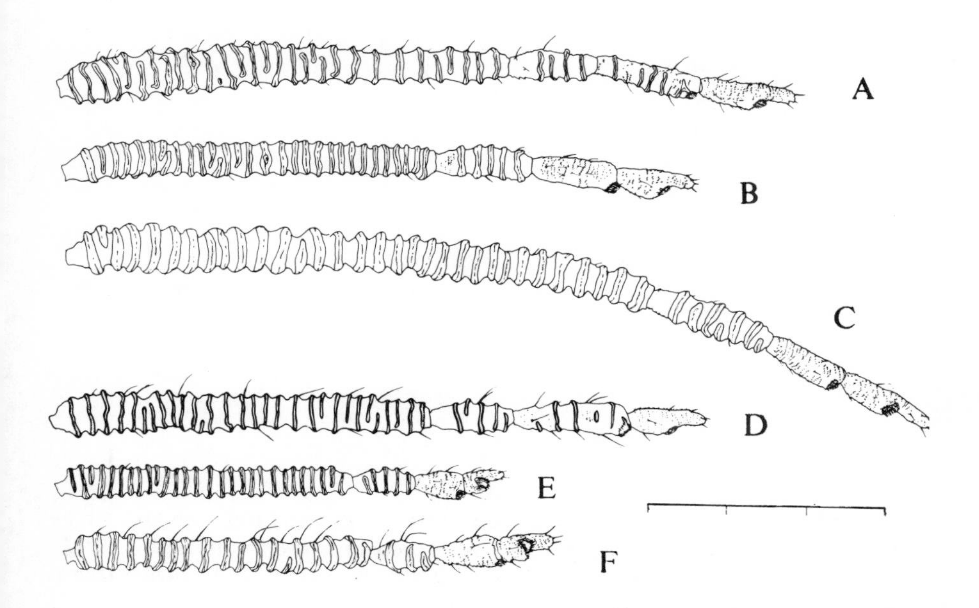

Fig. 126. Antennal segments III–VI of al. viv. (A–C: fundatrigeniae, D–F: sexuparae) of A & D: *Eriosoma (Schizoneura) patchiae* (Börn. & Bl.); B & E: *E. (S.) anncharlotteae* Dan.; C & F: *E. (S.) ulmi* (L.) (After Danielsson). (Scale 0.3 mm).

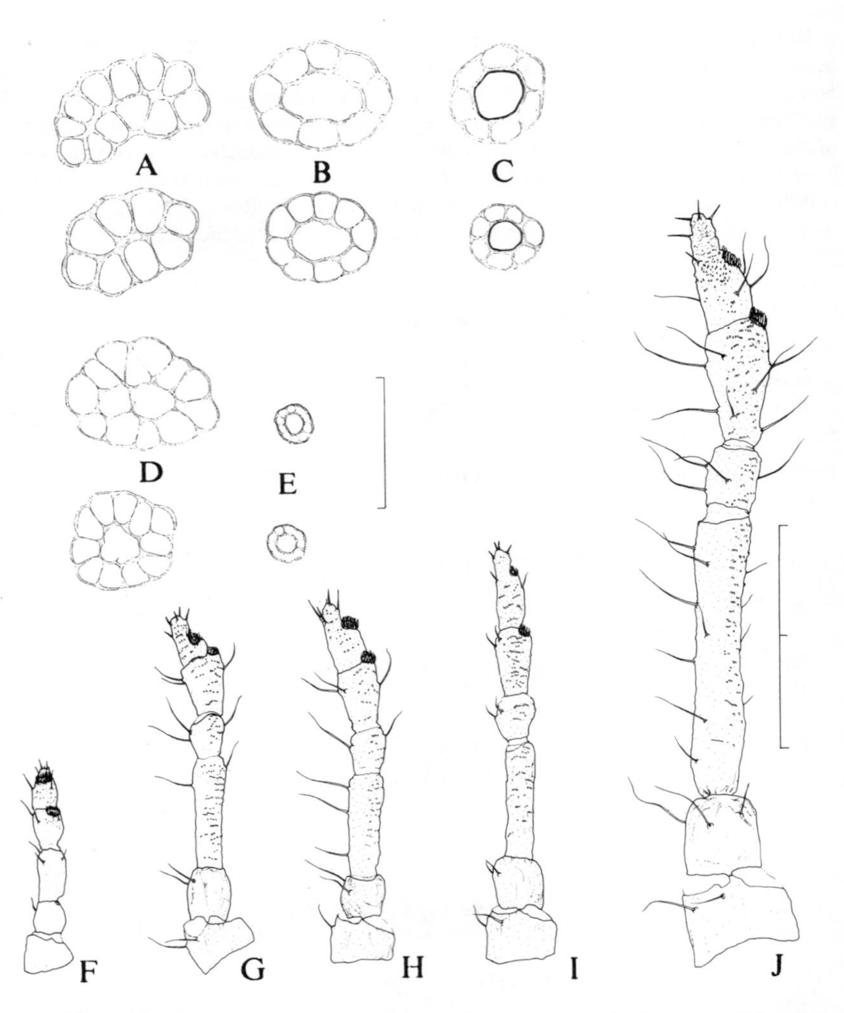

Fig. 127 A–E: Marginal (above) and spinal (below) wax gland plates on abd. tergite VI of apt. viv. on secondary hosts of A: *Eriosoma lanigerum* (Hausm.); B: *E. (Schizoneura) anncharlotteae* Dan.; C: *E. (S.) ulmi* (L.); D: *E. (S.) patchiae* (Börn. & Bl.); E: *E. (S.) sorbiradicis* Dan. (Scale 0.1 mm).

Fig. 127 F–J: Antennae of apt. viv. on secondary hosts of F: *Eriosoma (Schizoneura) sorbiradicis;* G: *E. (S.) anncharlotteae;* H: *E. (S.) ulmi;* I: *E. lanigerum;* J: *E. (S.) patchiae.* (After Danielsson). (Scale 0.2 mm).

distinct. First instar nymphs of apterae on secondary hosts with spinulose tarsi and normal dorsoapical hairs on second tarsal segments.

The subgenus consists of 11–12 species in the holarctic region; one of them is introduced into the southern hemisphere. Host-alternation takes place between *Ulmus* (I) and roots of various plants (II).

16. ***Eriosoma (Schizoneura) anncharlotteae*** Danielsson, 1979.
Plate 2: 5. Figs. 122 D, 125, 126 B & E, 127 B & G.

Eriosoma (Schizoneura) anncharlotteae Danielsson, 1979: 195.

Fundatrix. Dark green to bluish grey. Antennae usually 6-segmented, rarely 5-segmented. Very much like *ulmi,* but ant. segm. III a little shorter than IV+V+VI, longer than IV+V; VI about 0.6–0.7 × V. Longest hair on ant. segm. III 0.9–1.3 × IIIbd. Rostrum reaching to middle coxae; apical segm. 1.5–1.9 × 2sht. Siphuncular pores absent. 2.4–3.1 mm.

Apterous (alatiform) viviparous female of later generations on primary host (apterous fundatrigenia). Dark bluish grey. Body more elongate than in fundatrix. Eyes multifacetted. Antennae 6-segmented, longer than in fundatrix, about 0.4 × body. Rostrum reaching just past middle coxae; apical segm. 0.9–1.0 × 2sht. Siphuncular pores present. 2.0–2.2 mm.

Alate viviparous female from primary host (alate fundatrigenia). Dark bluish grey. Antennae (Fig. 126 B) 0.4–0.5 × body; segm. III 1.45 × IV+V+VI; V a little shorter than IV, 1.1–1.3 × VI (Fig. 122 D); processus terminalis about 0.65 × VIa; secondary rhinaria: III: 27–33, IV: 3–8, V: 0, VI: 0. Apical segm. of rostrum 0.13–0.16 mm, 0.8–1.0 × 2sht., with 12–14 accessory hairs. 1.8–2.3 mm.

Apterous viviparous female on secondary host (virginopara). Yellow to light red. Wax glands rather similar to those of *ulmi,* but the central facet not defined by a distinct double line (Fig. 127 B). Antennae 6-segmented, 0.3 × body; segm. III shorter than IV+V+VI, about as long as IV+V or a little longer. Apical segm. of rostrum 0.12–0.16 mm, 1.6–2.0 × 2sht., with 8–10 accessory hairs. Siphuncular pores large, placed on low, hairy cones. Cauda with 2 hairs. 1.1–1.9 mm.

Alate viviparous female from secondary host (sexupara). Wax gland plates rather small, but distinct. Ant. segm. III almost twice as long as IV+V+VI; V a little shorter than IV, about 1.5 × VI; secondary rhinaria: III: 23–33, IV: 3–6, V: 0, VI: 0 (Fig. 126 E). Apical segm. of rostrum 1.2–1.6 × 2sht. 1.6–2.2 mm.

Distribution. The species is only known from Sk. (Lund and Löberöd) in Sweden.

Biology. The primary hosts are *Ulmus carpinifolia* and *U. procera,* the secondary host *Ribes alpinum.* The leaves of the primary hosts are curled downwards on both sides. They are often completely distorted and clustered together. There may be several fundatrices in a single gall, and they become adults at the end of May in Skåne (S Sweden).

The alate fundatrigeniae usually leave *Ulmus* at the beginning of July, but galls with alatae were found by Danielsson as late as September. The sexuparae leave the roots of *Ribes alpinum* from mid-September onwards.

17. ***Eriosoma (Schizoneura) lanuginosum*** (Hartig, 1839)
Figs. 128, 129.

Schizoneura lanuginosa Hartig, 1839: 367. – Survey: 192.

Fundatrix. Bluish black, wax-powdered. Antennae 5-segmented. Body shape and other characters as in the other *E.* spp. 3.1–3.4 mm.

Apterous viviparous female of 2nd generation on primary host (apterous fundatrigenia). Olive-green, covered with wax; legs black. Wax gland plates distinct; the marginal plates with three very distinct central facets. Body not as thick as in the fundatrix. Antennae 6-segmented, about 0.3 × body. Rostrum reaching middle coxae. Siphuncular pores present.

Alate viviparous female from primary host (alate fundatrigenia). Almost black, covered with wax. Antennae 6-segmented; segm. III longer than IV+V+VI; V as long as or a little shorter than IV, much longer than VI; processus terminalis about 0.3 × VIa; secondary rhinaria: III: 28–35, IV: 7–11, V: 5–16, VI: 0–1. Rostrum hardly reaching middle coxae. Apical segm. of rostrum about 0.19–0.21 mm, about 1.3 × 2sht, with about 10–12 accessory hairs. Siphuncular pores present. 1.3–1.9 mm.

Apterous viviparous female on secondary host (virginogenia). Whitish yellow or reddish, covered with wax. Antennae 6-segmented, about 0.25 × body; segm. III shorter than IV+V; III and IV sometimes fused so that the antenna then is 5-segmented. Siphuncular pores on low cones. About 2 mm.

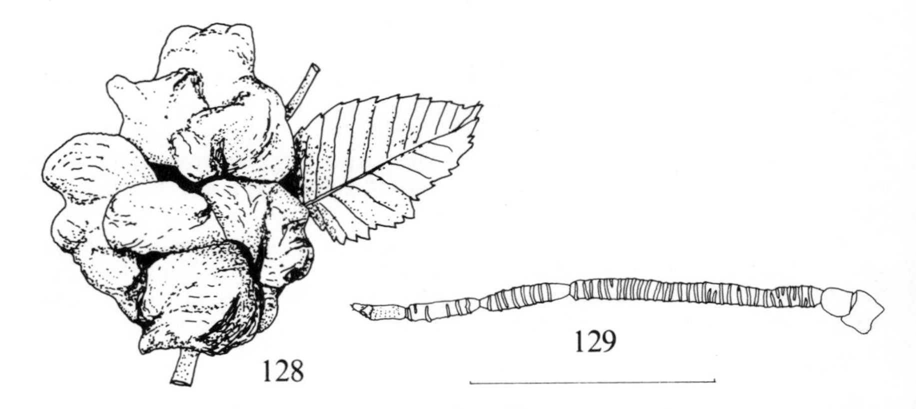

Figs 128, 129. *Eriosoma (Schizoneura) lanuginosum* (Hartig). – 128: galls on *Ulmus*. (Drawn after photo in Marchal); 129: antenna of alate fundatrigenia. (After Tullgren, redrawn). (Scale 0.5 mm for 129).

Alate viviparous female from secondary host (sexupara). Wax gland plates more distinctly visible than in the alate fundatrigenia. Processus terminalis about 0.5 × VIa. Secondary rhinaria: III: 16–27, IV: 3–6, V: 2–5, VI: 0.

Distribution. A few records of galls are given from Denmark (Jutland, Zealand) (Henriksen 1944); records from Sweden cannot be verified (Thomson 1862: no localities given; Danielsson 1979: material from Vg. was determined by Wahlgren, but the identification is doubtful); not in Norway or Finland. – Cosmopolitan; in Europe known from Great Britain, the Netherlands, Germany (rare in N Germany, common in S Germany), Poland (common), Russia (south of the forest zone), France, Switzerland, Hungary, Spain, Portugal, Yugoslavia, Bulgaria, W Asia, W Siberia, and C Asia; N Africa and S Africa; SE Australia, Tasmania, New Zealand; also known from the Pacific region of the USA and Canada and from some eastern states.

Biology. Closed, blister-shaped galls as large as walnuts or apples are produced in spring on *Ulmus (U. foliacea, U. suberosa, U. densa,* seldom *U. scabra).* The galls are light green, but become brown when old and are placed directly on thin branches. Some of the fundatrigeniae are apterous, others are alate and may belong to the first or second generation of fundatrigeniae. They fly to the thin roots of the secondary hosts which are members of the plant family Pomaceae *(Pyrus, Cydonia).*

18. ***Eriosoma (Schizoneura) patchiae*** (Börner & Blunck, 1916)
Figs. 122 C, 126 A & D, 127 D & J, 130.

Schizoneura patchiae Börner & Blunck, 1916: 30, 36. – Survey: 192.

Fundatrix. Green to dark bluish grey. Antennae usually 6-segmented, but segm. III and IV sometimes not distinctly separated; III longer than IV+V+VI; processus terminalis about 0.5 × VIa. Longest hair on ant. segm. III about 2 × IIIbd. Rostrum reaching to middle coxae; apical segm. 1.8–2.1 × 2sht. Siphuncular pores absent. 2.3–2.5 mm.

Apterous (alatiform) viviparous female of later generations on primary host (apterous fundatrigenia). Yellowish green, sometimes a little reddish, covered with wax. Glandular areas indistinct, but large, extending from side to side on posterior abd. tergites. Eyes with more than 3 ommatidia. Antennae as in fundatrix, but relatively longer, 0.4–0.5 × body. Apical segm. of rostrum 1.3–1.5 × 2sht. Large siphuncular pores present. 1.7–2.3 mm.

Alate viviparous female from primary host (alate fundatrigenia). Abdomen green or brownish, with narrow cross bands. Antennae (Fig. 126 A) 0.4–0.5 × body; segm. III 1.35–1.45 × IV+V+VI; V longer than IV, usually about 1.3 × VI (Fig. 122 C); processus terminalis 0.4 × VIa; secondary rhinaria: III: 18–35, IV: 2–7, V: 1–7, VI: 0. Rostrum reaching past middle coxae; apical segm. 0.23–0.25 mm, 1.1–1.5 × 2sht., with 19–23 accessory hairs. 2.1–2.5 mm.

Apterous viviparous female on secondary host (virginopara). Spinal wax gland plates with a central facet surrounded by 4–7 facets in a single row; marginal plates larger, oval, with 8 or more facets in the ring (Fig. 127 D). Antennae 6-segmented, about 0.3 ×

Fig. 130. Shoot of *Ulmus* with leaves curled by *Eriosoma (Schizoneura) patchiae* (Börn. & Bl.). (After Shaposhnikov, redrawn).

body; segm. III a little shorter than IV+V+VI, much longer than IV+V. Apical segm. of rostrum 0.22–0.24 mm, 1.8–2.0 × 2sht., with 17–26 accessory hairs. Siphuncular pores present. Cauda with 1–2 hairs. 1.9–2.7 mm.

Alate viviparous female from secondary host (sexupara). Wax glands distinct; the central facet often with indication of a subdivision into smaller facets; the surrounding facets in a single or double row. Ant. segm. III almost twice as long as IV+V, about 1.4 × IV+V+VI; IV and VI about equally long, V longer; secondary rhinaria: III: 22–30, IV: 3–5, V: 3–6, VI: 0 (Fig. 126 D). 2.0–2.3 mm.

Distribution. In Denmark rather common (Jutland, Zealand); in Sweden recorded from Sk. and Vg.; in Norway known from AK (trap); not in Finland. – Great Britain, Germany (but not in N Germany), Poland, Russia, Hungary; W Siberia, C Asia.

Biology. Fundatrices and 2–3 generations of apterous and alate fundatrigeniae develop on *Ulmus,* especially *U. glabra.* The species causes shortening of shoots and curling of leaves. Colonies can be observed on *Ulmus* through the whole summer. The species is dioecious, but the host-alternation is apparently facultative. Marchal (1933) found that sexuales can be born by alatae which are present on *Ulmus* as nymphs. Virtual sexuparae with more distinct wax glands than in alate fundatrigeniae arrive in autumn at *Ulmus* from secondary hosts. The secondary hosts are *Senecio* (Eastop 1961, Parker 1976), especially *S. jacobaea,* and *Cineraria* (Simmonds 1956). In Scandinavia, this species has only been found on the primary host.

19. ***Eriosoma (Schizoneura) sorbiradicis*** Danielsson, 1979
Figs. 127 E & F, 131.

Eriosoma (Schizoneura) sorbiradicis Danielsson, 1979: 206.

Apterous viviparous female on secondary host (virginogenia). Yellowish white. Antennae 5- or 6-segmented (the border between segm. III and IV is indistinct or absent) (Fig. 127 F); 0.11–0.20 × body; ultimate segm. a little shorter than penultimate segm. Apical segm. of rostrum 0.07–0.10 mm, 1.5–2.0 × 2sht., with 5–7 accessory hairs. Wax gland plates (Fig. 127 E) rather small, but distinct, of the same type as in *ulmi*. Siphuncular pores rather large, placed on low, hairy cones. Cauda with 3–5 hairs. 0.7–1.4 mm.

Distribution. The species is only known from Sk. and Upl. in Sweden.

Biology. The virginogeniae have been found on roots of *Sorbus aucuparia* in summer. The other morphs are unknown.

Note. The species is rather similar to *lanuginosum,* which also lives on Pomaceae in summer, but the body is smaller, and the antennae, especially segm. III, are relatively shorter.

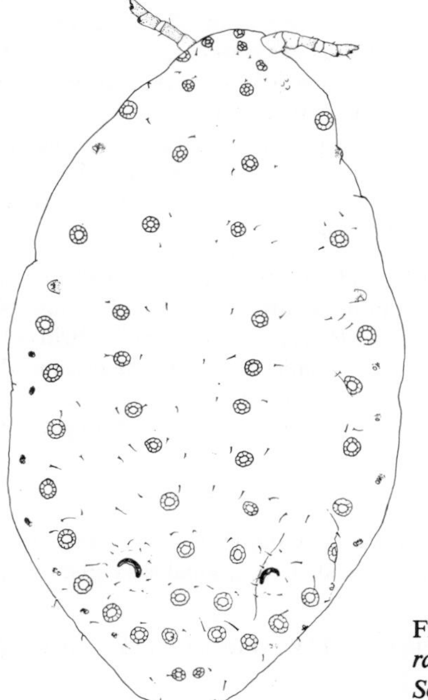

Fig. 131. *Eriosoma (Schizoneura) sorbiradicis* Dan., apt. viv. from root of *Sorbus aucuparia.* (After Danielsson). (Scale 0.3 mm).

20. ***Eriosoma (Schizoneura) ulmi*** (Linné, 1758)
Plate 1: 5. Figs. 30 Q, 122 B, 126 C & F, 127 C & H, 132–135.

Chermes ulmi Linné, 1758: 455. – Survey: 192.

Fundatrix. Dark green, almost black, covered with wax. Body broadly oval, vaulted. Wax gland plates present, but rather indistinct. Antennae 6-segmented, 0.18–0.20 × body; segm. III longer than IV+V+VI; V about as long as IV, about 1.3 × VI; processus terminalis about 0.25 × VIa. Longest hair on ant. segm. III 1.0–1.5 × IIIbd. Rostrum hardly reaching middle coxae; apical segm. 1.33 × 2sht. Siphuncular pores absent. 2.8–3.8 mm.

Alate viviparous female from primary host (fundatrigenia). Abdomen dark brown, with irregular, dark, dorsal cross bands. Covered with wax; glands hardly visible. Antennae (Fig. 126 C) 6-segmented, about 0.5 × body; segm. III about twice as long as IV+V+VI; V shorter than IV, as long as or up to 1.15 × VI (Fig. 122 B); processus terminalis 0.4–0.5 × VIa; secondary rhinaria: III: 30–46, IV: 5–9, V: 0,VI: 0. Rostrum hardly reaching to middle coxae; apical segm. shorter than 0.20 mm, usually 0.16–0.17 mm, 0.8 × 2sht., with about 15–18 accessory hairs. Siphuncular pores large, placed on very low, rather dark, hairy cones. 2.0–3.0 mm.

Apterous viviparous female on secondary host (virginogenia). Yellowish or reddish yellow, with wax threads produced by gland plates consisting of a large, central facet, defined by a distinct double line, and a ring of smaller facets (Fig. 127 C). Antennae 6-segmented, about 0.2 × body; segm. III about as long as IV+V. Apical segm. of rostrum shorter than 0.20 mm. Siphuncular pores present. About 1.7 mm.

Alate viviparous female from secondary host (sexupara). Wax gland plates much more distinct than in the alate fundatrigenia, the marginal plates larger than the spinal plates. Antennae (Fig. 126 F) shorter than in the alate fundatrigenia. Secondary rhinaria: III: 18–22(–27), IV: 3–4, V: 0, VI: 0. 1.7–1.9 mm.

Distribution. Very common and widespread in Denmark; in Sweden common, occurring north to Med.; in Norway common, occurring north to Nsy and NTi; in Finland also common, north to Om and Sb. – All over Europe from the British Isles to Russia, south to Portugal and Yugoslavia, very common in N Germany and Poland, also along the Baltic coast, and in the Baltic region of Russia; recorded from the Faroes and Iceland. Widespread also in Asia: W Siberia, C Asia, Mongolia, China, Turkey, Lebanon, and Iraq.

Fig. 132. Fundatrix of *Eriosoma (Schizoneura) ulmi* (L.) – A: body in dorsal view. (Scale 1 mm); B: wax gland plate on head; C: marginal wax glands on abd. segm VII.

Figs. 133, 134. Same, alate fundatrigenia. – 133: antenna; 134: wings. (Scales 0.1 mm for 133, 1 mm for 134).

Fig. 135. Same, gall on *Ulmus*. (After Shaposhnikov, redrawn).

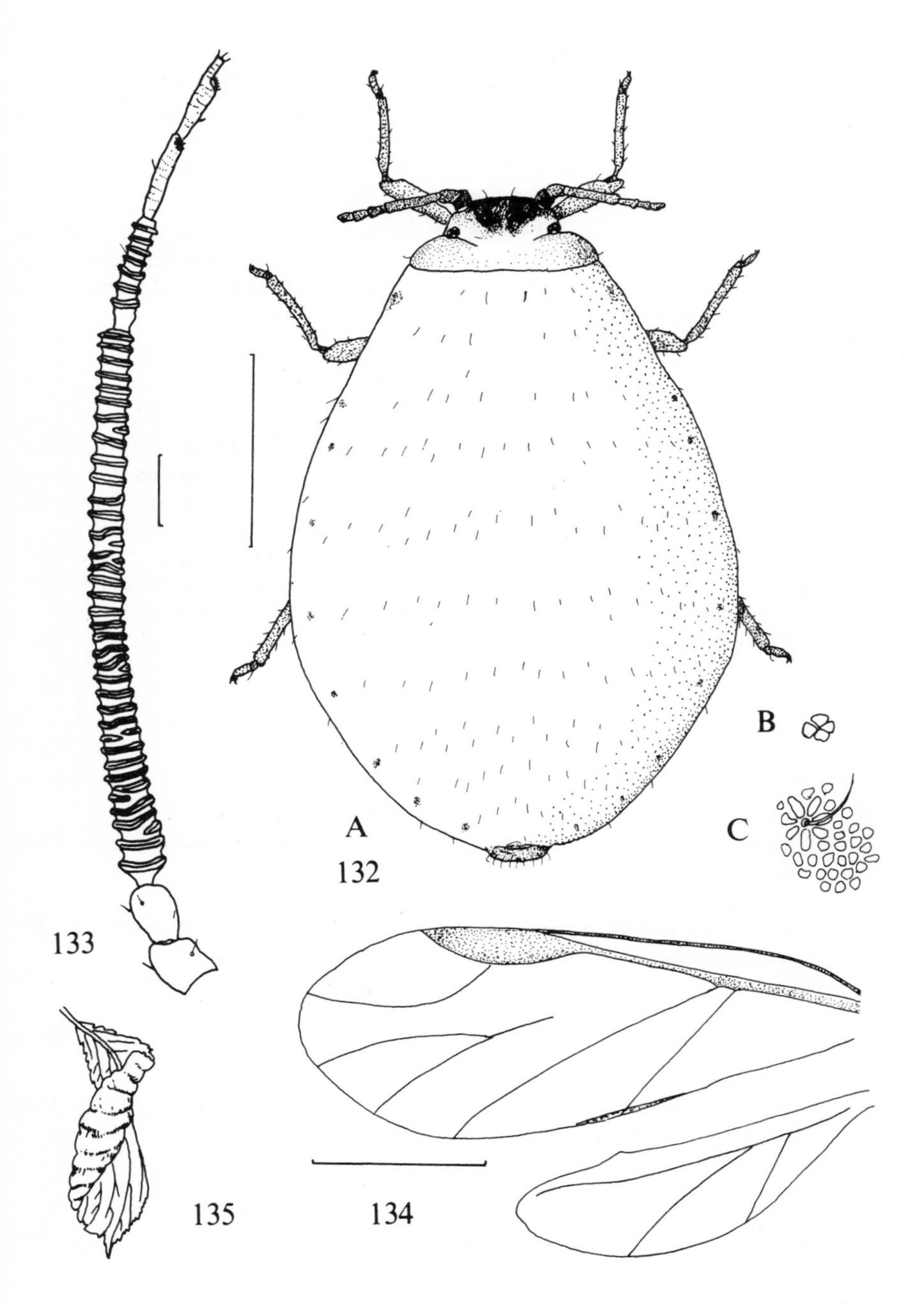
A
132
B
C
133
135
134

Biology. The species produces an open gall on *Ulmus*, the primary host, in spring by rolling the leaf from one side towards the underside (Fig. 135). The roll becomes yellowish. In Denmark the fundatrix becomes an adult in May. The alate fundatrigeniae fly to the roots of *Ribes (Ribes rubrum, R. nigrum, R. uva-crispa* a. o.), in Denmark in June.

Note. Danielsson (1979) found that his material of *E. (S.) ulmi* could be separated into two forms with different types of wax glands and some other morphological details. They are apparently equally common in Sweden and may be different species. Danielsson will publish the results of his studies in a future paper.

TRIBE TETRANEURINI

Media of fore wing unbranched or with one fork. Hind wing with one or two oblique veins. Ant. segm. III of alatae as long as IV+V+VI or shorter. Siphuncular pores sometimes absent in some morphs, including alatae. Apterae on secondary hosts with 4–6 longitudinal rows of wax gland plates and often with only one tarsal segment.

More or less closed galls, which are local swellings of the upper surface of the leaves, are produced on the primary host, *Ulmus*. The secondary hosts are herbs belonging to various families. On these the aphids feed on the roots and are visited by ants.

Three genera occur in the region.

Key to genera of Tetraneurini

Galls on *Ulmus*

1 Usually placed at the base of the mid rib of the leaf; globular, short-haired (Fig. 143). .... *Kaltenbachiella* Schouteden (p. 133)
– Placed outside the mid rib of the leaf; not globular, without hairs (in Scandinavian species). 2
2 (1) Bean-shaped, with short stalk (Fig. 149). .... *Tetraneura* Hartig (p. 135)
– Cockscomb-shaped, compressed, without stalk, broad at base (Fig. 140). .... *Colopha* Monell (p. 131)

Alate viviparous females

1 Hind wing with two oblique veins (Fig. 141). Ant. segm. VI with 2–10 secondary rhinaria. .... *Kaltenbachiella* Schouteden (p. 133)
– Hind wing with only one oblique vein (Fig. 136). Ant. segm. VI with 0–2 secondary rhinaria. 2
2 (1) Ant. segm. III as long as segm. IV + V, but shorter than IV+V+VI. Ant. segm. VI usually without secondary rhinaria, seldom with one. Media of fore wing not branched. .... *Tetraneura* Hartig (p. 135)

– Ant. segm. III longer than segm. IV+V, as long as IV+V+VI. Ant. segm. VI with 0–2 secondary rhinaria. Media of fore wing with one fork. .... *Colopha* Monell (p. 131)

Apterous viviparous females on secondary hosts

1 Hind tarsi usually 2-segmented. Antennae usually 4-segmented. On Labiatae. .... *Kaltenbachiella* Schouteden (p. 133)
– Hind tarsi with only one segment. Antennae often 5- or 6-segmented. On monocotyledones. 2
2 (1) Anus surrounded by long hairs (Fig. 148). On grasses. .. *Tetraneura* Hartig (p. 135)
– Anus not surrounded by long hairs. On *Carex*. .... *Colopha* Monell (p. 131)

Genus *Colopha* Monell, 1877

Colopha Monell, 1877: 102.
Type-species: *Byrsocrypta ulmicola* Fitch, 1859.
Survey: 160.

One species in Scandinavia, 8 species in the world.

21. ***Colopha compressa*** (Koch, 1856)
Figs. 136–140.

Schizoneura compressa Koch, 1856: 267. – Survey: 160.

Fundatrix. Yellow or yellowish green, weakly wax powdered. Eyes red. Antennae and legs brownish. Body elongate, oval or oviform, vaulted, with only a few very short hairs. Wax gland plates consisting of 1–2 central facets surrounded by a ring of facets may be present, often 4 plates on prothorax, 4 on mesothorax, 6 on metathorax and each of the abd. segm. I–VI, 4 on VII, and 2 on VIII. Antennae 4- or 5-segmented, about 0.12 × body. Primary rhinaria swollen, not surrounded by short hairs; accessory rhinaria on distal ant. segm. surrounded by short hairs. Small siphuncular pores present or absent. Tarsi 1- or 2-segmented; if 2-segmented, then first tarsal segm. with tuberculate distal end. 1.2–1.6 mm.

Alate viviparous female from primary host (fundatrigenia). Abdomen green. Body with a few fine hairs. Without distinct wax gland plates. Antennae 6-segmented (but the border between segm. III and IV sometimes indistinct), about 0.25 × body; segm. III as long as IV+V+VI; V a little longer than VI. Primary rhinarium on VI of irregular shape, swollen(blister-shaped), not surrounded by short hairs; accessory rhinaria on the same segm. small, surrounded by short hairs (Fig. 138). Primary rhinarium on V and secondary rhinaria narrow, ringlike, not surrounded by hairs. Ant. segm. III with 11–15 secondary rhinaria, IV with 2–4, V with 1–3, VI with 0–2. Rostrum reaches to the middle of mesosternum. Tarsi 2-segmented; first tarsal segm. with 3-2-2 hairs. Media of fore wing with one fork. Hind wing with only one distinct oblique vein. Siphuncular pores present or absent (present in specimens studied from Denmark and the Netherlands). About 1.4–1.8 mm.

Apterous viviparous female on secondary host (virginogenia). Dark yellowish, with wax powder and wax flock. Wax gland plates consisting of one or more central facets, surrounded by a ring of 8–12 facets, present, on head 8–10 plates, on prothorax 1 spinal pair of plates and 1 marginal pair, on meso- and metathorax and abd. segm. I–III 1 spinal, 1 pleural, and 1 marginal pair, on IV–VI 1 spinal and 1 marginal pair. Antennae

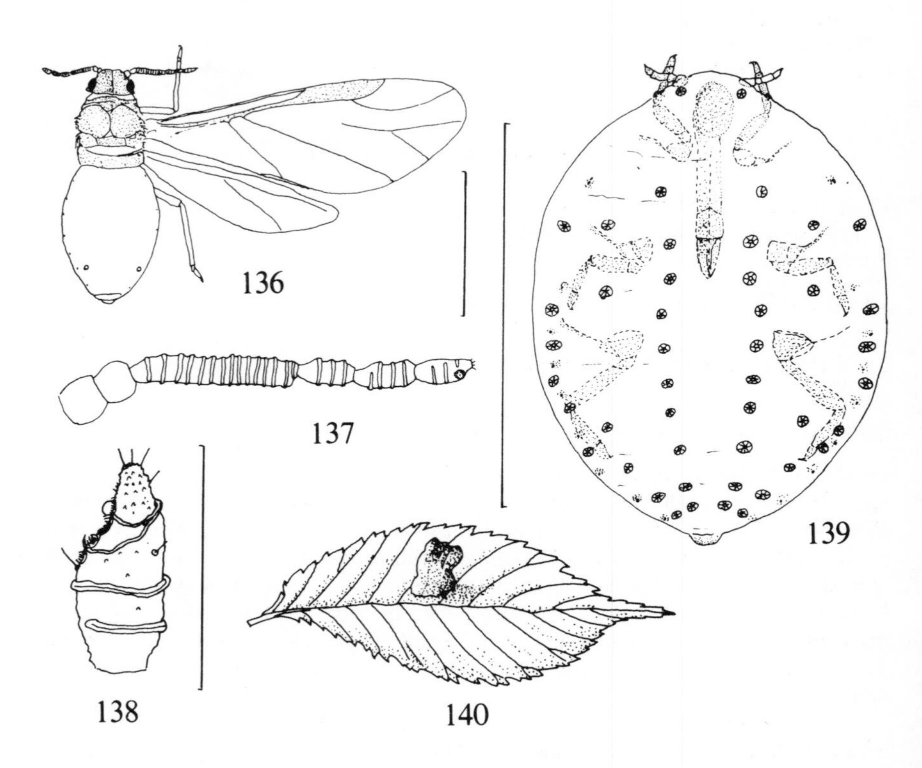

Figs. 136–138. *Colopha compressa* (Koch), alate fundatrigenia. – 136: body and wings; 137: antenna; 138: ant. segm. VI. (136 & 137 after Mordvilko, redrawn). (Scales 1 mm for 136, 0.1 mm for 138).

Fig. 139. Same, virginogenia from *Carex,* dorsal view (rostrum and legs, which are visible only from the underside, are shown in dotted lines). (Scale 1 mm).

Fig. 140. Same, gall on upper surface of leaf of *Ulmus.* (After Mordvilko, redrawn).

4- or 5-segmented, about 0.10–0.12 × body; primary rhinaria surrounded by short hairs. Tarsi 1-segmented. Very small siphuncular pores present or absent. Anus not surrounded by long hairs. 1.0–1.4 mm.

Alate viviparous female from secondary host (sexupara). Very much like fundatrigenia, but primary rhinaria surrounded by rings of short hairs. Ant. segm. III with about 11 secondary rhinaria, IV with 3–4, V with 2–3, VI with 0–2.

Distribution. In Denmark found in Skive (NWJ); in Sweden found in Sk., Sm., and Upl.; not recorded from Norway; in Finland found in Ab, N, Ta, and Sb. – Widespread in Europe, from Scandinavia south to the Mediterranean; known from Iceland, N Germany (common), and the Baltic coast of Poland. The species occurs east to Russia and Turkey and has been introduced in towns of Siberia.

Biology. The species is holocyclic and dioecious. The fundatrices produce compressed, cockscomb-shaped, smooth, yellowish or reddish galls (Fig. 140) along with the mid ribs of leaves of *Ulmus laevis* (also recorded from *U. scabra* and *U. effusa),* the primary host. On the secondary hosts, species of *Carex,* the aphids are found on the roots, sometimes in ants' nests (e.g. of *Lasius flavus).*

Note. According to Zwölfer (1957) Tullgren's Swedish samples are somewhat different from C European samples, but conspecific. Tullgren (1925) informed that fundatrices have siphuncular pores, 5-segmented antennae, and 2-segmented tarsi, while fundatrices from Germany are without siphuncular pores and have 4-segmented antennae and 1-segmented tarsi. Swedish and Russian apterae from secondary hosts have siphuncular pores, whereas these are absent from specimens from Denmark and C Europe.

Genus *Kaltenbachiella* Schouteden, 1906

Kaltenbachiella Schouteden, 1906: 194.
Type-species: *Kaltenbachiella menthae* Schouteden, 1906.
= *Byrsocrypta pallida* Haliday, 1838.
Survey: 232.

One species in Scandinavia, 4 species in the world.

22. ***Kaltenbachiella pallida*** (Haliday, 1838)
Figs. 141–143.

Byrsocrypta pallida Haliday, 1838: 189. – Survey: 232.

Fundatrix. Yellowish. Body elongate, oval. Wax gland plates absent. Body with very few short hairs. Antennae 4-segmented, about 0.15 × body; segm. III about 2 × IV. Primary rhinaria small, round, without distinct rings of surrounding short hairs. Rostrum reaching middle coxae. Hind tarsi 2-segmented, fore and middle tarsi usually 1-segmented. Siphuncular pores absent. About 2 mm.

Alate viviparous female from primary host (fundatrigenia). Abdomen dark green. Wax gland plates absent. Antennae 6-segmented, about 0.4 × body; segm. III with 20–24 transverse narrow secondary rhinaria, IV with 6–7, V with 5–8, VI with 7–10; segm. III slightly shorter or slightly longer than IV + V; primary rhinaria not surrounded by short hairs. Media of fore wing usually unbranched, seldom with one fork; cubitus 1a and 1b leave the main vein in separate points. Hind wing with two oblique veins leaving the main vein in widely separate points. All tarsi 2-segmented. Siphuncular pores absent. About 2.1 mm.

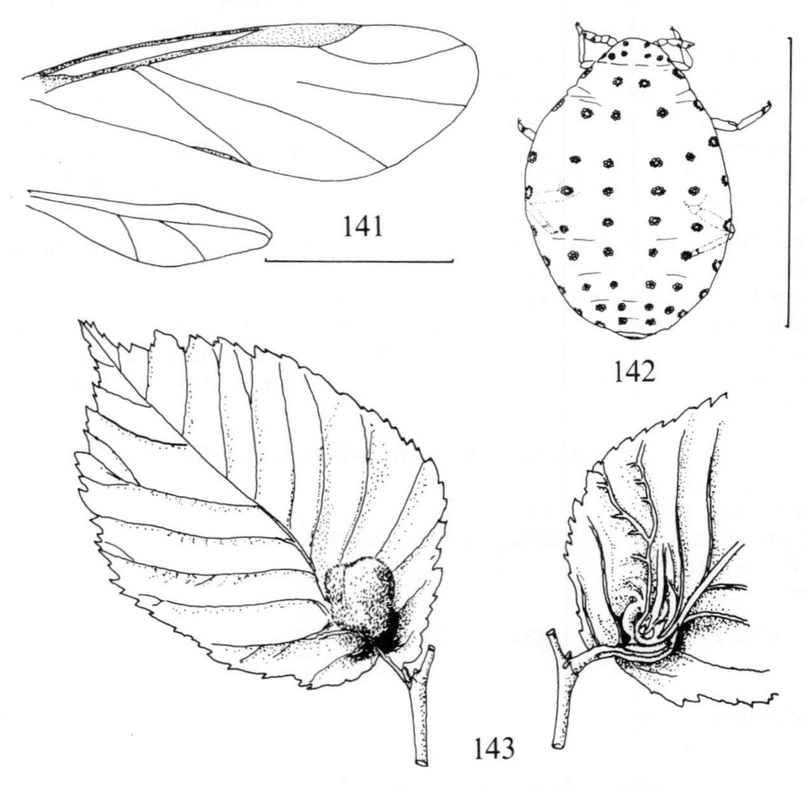

Fig. 141. *Kaltenbachiella pallida* (Hal.), wings of fundatrigenia. (Scale 1 mm).

Fig. 142. Same, apt. viv. from root of *Mentha* (here with one-segmented fore tarsi). (After Mordvilko, redrawn). (Scale 1 mm).

Fig. 143. Same, gall on leaf of *Ulmus,* upper surface to the left, lower surface to the right. (After Marchal, redrawn).

Apterous viviparous female on secondary host (virginogenia). Yellowish white; as a nymph pale orange yellow; with wax flock. Body subcircular or oviform, vaulted. Wax gland plates consisting of one usually elongate central field surrounded by a ring of roundish facets present, the marginal plates being larger than the dorsal ones; head with 8–10 plates, prothorax with 1 spinal pair of plates and 1 marginal pair, meso- and metathorax and abd. segm. I–VI with 1 spinal pair, 1 pleural pair, and 1 marginal pair, VII with 1 spinal pair and 1 marginal pair, and VIII with 2 small plates. Antennae 4-, seldom 5-segmented, 0.12–0.15 × body; segments equally long, or III longer than the others; primary rhinaria round or oval, surrounded by short hairs. Fore and middle tarsi usually 1-segmented, hind tarsi 2-segmented (but the limit between the segments may be indistinct). Siphuncular pores absent. Anus not surrounded by long hairs. 0.9–1.3 mm.

Alate viviparous female from secondary host (sexupara). Abdomen olive green, faintly wax powdered. Very much like fundatrigenia, but the primary rhinarium on ant. segm. V of the same narrow, ringlike shape as the secondary rhinaria; ant. segm. III with 10–12 secondary rhinaria, IV with 3–4, V with 1–2, VI with 2–3; primary rhinarium on VI subcircular or oval. Tarsi usually 2-segmented. Siphuncular pores absent. About 1.0 mm.

Distribution. In Denmark rare, known from EJ and NEZ; in Sweden found in Sk., Gtl., Öl., Ög., and Vg.; not recorded from Norway; in Finland only known from Ta. – The species occurs all over Europe, from S Scandinavia south to Spain and Portugal, from Britain and France east to Russia. Also W Siberia, C Asia, N Africa, Turkey, Lebanon, and Iraq.

Biology. The species is host-alternating between *Ulmus (U. campestris, U. scabra)*, the primary host, and subterranean parts of Labiatae (e.g. *Mentha, Thymus, Origanum*, and *Galeopsis*), the secondary hosts. Fundatrices produce in spring globular, closed, pale, densely short-haired galls on the leaves of *Ulmus*, projecting both on the upper- and underside of the leaf (Fig. 143). When the alate fundatrigeniae are ready to leave the primary host, the gall achieves a star-like opening in the top. Anholocyclic overwintering on the roots of the secondary hosts may possibly take place in C Europe.

Genus *Tetraneura* Hartig, 1841

Tetraneura Hartig, 1841: 366.

Type-species: *Aphis ulmi* Linné, 1758.

Survey: 418.

All legs of apterae with one tarsal segment. Antennae 3- or 4-segmented in fundatrices, 5- or 6-segmented in apterous females on secondary hosts, 6-segmented (seldom 5-segmented) in alate females. Forewing with unbranched media (Fig. 145). Hind wing with only one oblique vein. Siphuncular pores present or absent. Wax gland plates present or absent.

The genus is palaearctic, but now spread all over the world. There are 20 species in *Tetraneura* s. lat., 12 species in *T.* s. str.; two species are recorded from Scandinavia, but they may be varieties of the same species.

Ulmus is the primary host, grasses (Gramineae) are the secondary hosts. The aphids are visited by ants on the secondary hosts.

Key to species of *Tetraneura*

Apterous viviparous females on secondary hosts (the other morphs of *longisetosa* are little known)

1 Abdominal hairs usually 0.03–0.04 mm. .... 24. *ulmi* (Linné)
– Abdominal hairs often longer than 0.04 mm. .... 23. *longisetosa* (Dahl)

23. ***Tetraneura longisetosa*** (Dahl, 1912)

Tycheoides longisetosa Dahl, 1912: 435. – Survey: 419.

Apterous viviparous female on seecondary host (virginopara). The average number of hairs is larger than in *ulmi* (but the range of variation is about the same), and their average length is also larger, the marginal hairs being up to 0.21 mm long. The ring of the marginal wax gland on tergite VII consists of one row of facets (in *ulmi* usually more than one row). Otherwise very similar to *ulmi.*

Distribution. In Scandinavia recorded only from Sweden (Upl.). – Outside Scandinavia only known from Germany (Dahl 1912, Börner 1952, Zwölfer 1957).

Biology. The species was found on roots of grasses *(Brachypodium, Deschampsia,* and *Festuca),* mostly in shaded localities (Zwölfer 1957). Börner collected sexuparae from colonies on grass roots in autumn. The primary host is probably *Ulmus,* but this needs verification.

Note. Hille Ris Lambers (1970) was of the opinion that it is identical with *ulmi.* Zwölfer (1957) regarded it as a separate species although the range of variation in hair length and in the morphology of the wax glands is about the same as in *ulmi,* and gave arguments based on a statistical analysis.

24. ***Tetraneura ulmi*** (Linné, 1758)
Plate 1: 6–7 and 2: 2. Figs. 144–149.

Aphis ulmi Linné, 1758: 451. – Survey: 419.

Fundatrix. Light olive-green. Antennae and legs brownish. Head and thorax dark and distinctly limited against the nearly globular abdomen. Wax gland plates absent. Body with only few short hairs. Antennae 3- or 4-segmented (the two distal segments may be separated or not), hardly 0.15 × body; primary rhinaria small, roundish. 1.9–2.1 mm.

Alate viviparous female from primary host (fundatrigenia). Head, thorax, antennae, and legs shining black. Abdominal dorsum greyish black, venter greyish green. Wax gland plates consisting of a large central facet and a ring of strongly sclerotized small facets may be present on head, thorax, and abdomen. Antennae 6-segmented, about 0.33 × body; segm. III with 8–17 secondary rhinaria, which are narrow and ringlike, covering 1/2–3/4 of the circumference of the segment, IV with 2–5, V with 4–8, VI with 0–1; primary rhinaria irregular, that on VI star-shaped; III as long as IV + V, VI shorter than V; processus terminalis about 0.25 × VIa. Rostrum reaches nearly to middle coxae; apical segm. shorter than 2sht. Tarsi 2-segmented. Cubitus 1a and 1b of fore wing leave main vein in almost the same point. Siphuncular pores present or absent. 1.8–2.6 mm.

Apterous viviparous female on secondary host (virginogenia). Orange yellow, whitish yellow, or red, sometimes dark red. Slightly wax powdered. Hairs variable; abdominal hairs usually 0.03–0.04 mm. With wax gland plates as in alatae, 3–4 pairs on head, 1 pair of spinal and 1 pair of marginal plates on prothorax, 1 pair of spinal, 1 pair of pleural, and 1 pair of marginal plates on meso- and metathorax and abd. segm. I–VI, 1 pair of spinal and 1 pair of marginal plates (Fig. 147) on VII. Antennae 5-segmented (rarely 6-

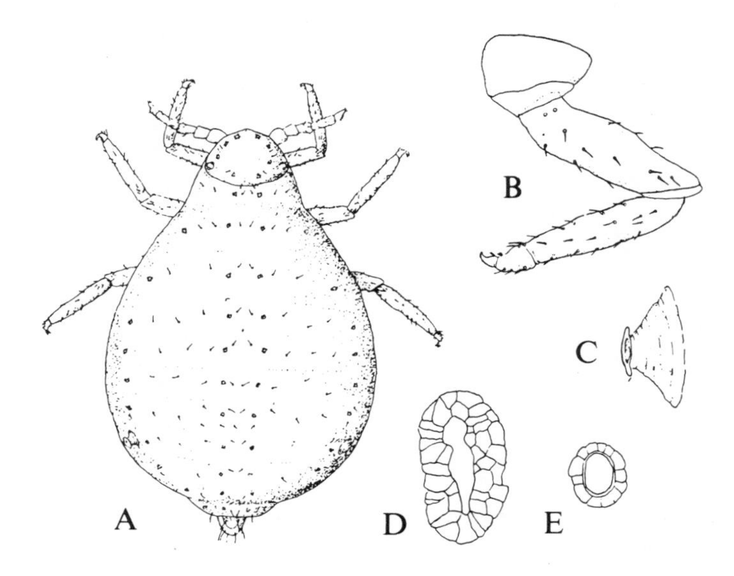

Fig. 144. *Tetraneura ulmi* (L.). – A: apt. viv. from grass-roots; B: leg showing the one-segmented tarsus; C: siphunculus; D & E: wax gland plates. (After Martelli, from Börner & Heinze).

segmented; some of these specimens have more than 3 ommatidia in the eye), 0.15–0.20 × body; primary rhinaria surrounded by short hairs. Rostrum nearly reaching to hind coxae. Tarsi 1-segmented. Siphuncular pores present on low, conical warts. Anus surrounded by long hairs (Fig. 148). 1.9–3.0 mm.

Alate viviparous female from secondary host (virginopara or sexupara). Similar to fundatrigenia, but ant. segm. III and IV sometimes fused, and the penultimate segm. (V in 6-segmented antennae, IV in 5-segmented antennae) with more (6–13) secondary

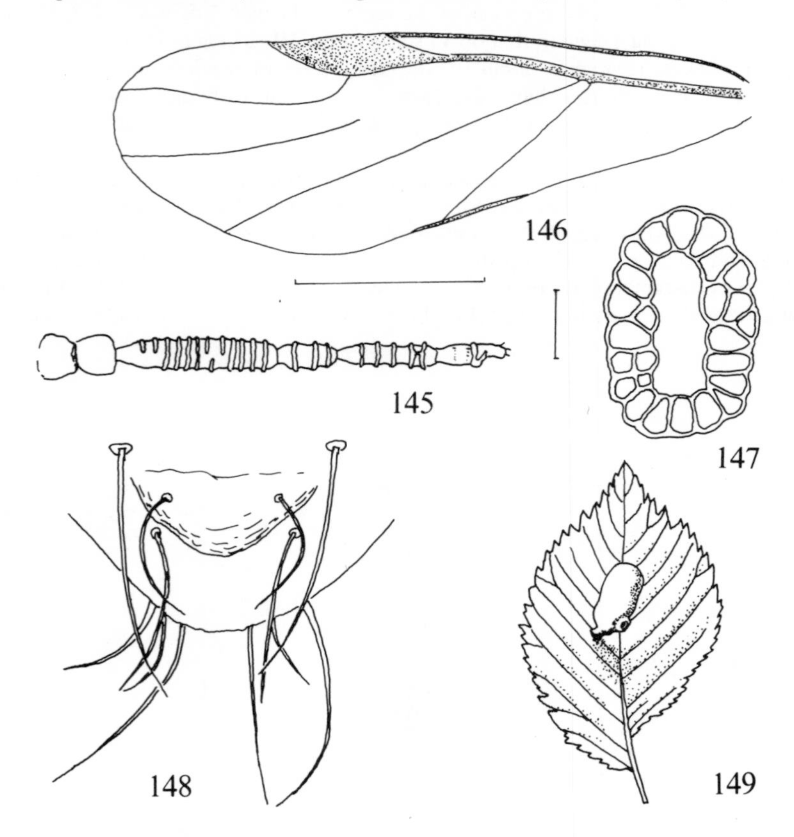

Figs. 145, 146. *Tetraneura ulmi* (L.), alate fundatrigenia. – 145: antenna; 146: fore wing. (Scale 1 mm).

Figs. 147, 148. Same, apt. viv. from grass-roots. – 147: marginal wax gland plate on abd. segm. VII. (Scale 0.1 mm); 148: posterior end of abdomen.

Fig. 149. Same, gall on upper surface of leaf of *Ulmus*.

rhinaria, and 2–3 times as long as the ultimate segm. Apical segm. of rostrum longer than 2sht. Wax gland plates always present, but varying in number. Siphuncular pores present.

Distribution. Widespread and rather common in Denmark, known from all districts, in some years very common, though never as common as *Eriosoma ulmi;* common in Sweden, known from Sk. in the South to T. Lpm. in the North; in Norway known from several districts north to Troms (TRi); rather common to common all over S Finland, north to St, Ta, and Sa. – Common and widespread all over Europe, incl. the Mediterranean, in Russia north to Leningrad; W Siberia, C Asia, Middle East (Iraq, Syria, Turkey) and Iran. Originally palaearctic; introduced into the Azores and N America.

Biology. The galls on *Ulmus* occur on the upper surface of the leaves. One leaf may carry several galls. They are bean-shaped, higher than wide, stalked, smooth and shining, green or yellowish (Fig. 149). The gall gets a basal opening on the side, when the fundatrigeniae are fully grown and ready to take off for grasses, which are the secondary hosts. The virginogeniae live on grass roots, often in ants' nests, and here the sexuparae are born. In autumn they fly to *Ulmus* to bear the apterous, dwarfish sexuales, which produce the overwintering eggs.

SUBFAMILY PEMPHIGINAE

Antennae of alate females 5- or 6-segmented. Secondary rhinaria transverse oval or narrow, often stripe-shaped, covering about half of the circumference of the segments. Media of fore wing unbranched or with one fork. Hind wing with two oblique veins which are not separated at bases or hardly so (Fig. 152). Tarsi two-segmented in all morphs (except in apterous virginogeniae of *Mimeuria*). Siphuncular pores present or absent, in apterae on secondary hosts always absent. Apterae on secondary hosts and generally also other morphs with honeycomb-like wax gland plates (Figs. 155, 162), often with white tuft of wax threads covering the posterior part or the whole body.

Most holocyclic species are host-alternating between deciduous trees, in most cases *Populus* (I), and roots of coniferous trees or roots or aerial parts of herbaceous (rarely woody) dicotyledones (II).

Keys to genera of Pemphiginae

Fundatrices

1 Body without wax gland plates (Fig. 154). .... *Pachypappa* Koch (p. 143)
– Body with wax gland plates (Figs. 176, 177, 188). 2
2 (1) Head with well developed wax gland plates. 3
– Head without wax gland plates. 7

3 (2) Primary rhinaria not surrounded by distinct rings of short hairs. On *Populus tremula.* .... *Gootiella* Tullgren (p. 151)
– Primary rhinaria surrounded by distinct rings of short hairs. Not on *Populus tremula.* 4
4 (3) Ant. segm. III with 4–6 hairs, the longest being distinctly longer than IIIbd. .... *Mimeuria* Börner (p. 163)
– Ant. segm. III with less than 4 hairs or more than 6, the longest being about as long as IIIbd. or shorter. 5
5 (4) Apical segment of rostrum without pale subapical zone (Fig. 165 B). On *Lonicera* or *Crataegus.* .... *Prociphilus* subg. *Stagona* Koch (p. 159)
– Apical segment of rostrum with at least a narrow pale subapical zone (Fig. 165 A). Not on *Lonicera* or *Crataegus.* 6
6 (5) Apical segment of rostrum as long as 2sht., with 4 or more accessory hairs. Ant. segm. III with 20 or more hairs shorter than or about as long IIIbd. On *Fraxinus.* .... *Prociphilus* Koch s. str. (p. 154)
– Apical segment of rostrum distinctly shorter than 2sht., with 0–2 accessory hairs. Ant. segm. III with 1–2 hairs, which are shorter than IIIbd. On *Populus nigra* and *P. italica.* .... *Thecabius* Koch (p. 166)
7 (2) Antenna 4-segmented (segm. III may be indistinctly subdivided into two segments). On *Populus nigra* and *P. italica.* ...... *Pemphigus* Hartig (p. 171)
– Antenna 5-segmented. On *Populus tremula.* .... *Pachypappella* Baker (p. 150)

Alate viviparous females from primary hosts (fundatrigeniae)

1 Media of fore wing with one fork. 2
– Media of fore wing unbranched. 4
2 (1) Primary rhinaria very large, star-shaped, surrounded by indistinct rings of short hairs (Fig. 163). .... *Gootiella* Tullgren (p. 151)
– Primary rhinaria not very large, nor star-shaped, but round or oval, surrounded by distinct rings of short hairs. 3
3 (2) Siphuncular pores large, diameter longer than 3 × IIIbd. *Pachypappella* Baker (p. 150)
– Siphuncular pores small, sometimes hardly visible, diameter shorter than 2.3 × IIIbd. .... *Pachypappa* Koch (p. 143)
4 (1) Ant. segm. VI without secondary rhinaria. 5
– Ant. segm. VI with secondary rhinaria. 7
5 (4) Secondary rhinaria without dotted borders. Head without wax gland plates. .... *Pemphigus* Hartig (partim) (p. 171)
– Secondary rhinaria with dotted borders. Head with or without distinct pale wax gland plates. 6
6 (5) Head with distinct wax gland plates. .... *Prociphilus* Koch (p. 153)
– Head without wax gland plates. .... *Mimeuria* Börner (p. 163)
(see also *Pachypappa warschavensis*,(p. 149)

7 (4) First tarsal segm. with 2 hairs on all legs. .... *Pemphigus* Hartig (partim) (p. 171)
– First tarsal segm. with 2–5 hairs, 3 or more hairs on at least some of the tarsi. .... *Thecabius* Koch (p. 166)

Apterous viviparous females on secondary hosts
(except *Pachypappella* and *Gootiella*)

1 Apical segm. of rostrum without distinct pale subapical zone (Fig. 165 B). 2
– Apical segm. of rostrum with distinct pale subapical zone (Fig. 165 A). 3
2 (1) Tarsi one-segmented. Processus terminalis vestigial, not projecting more than 0.01 mm beyond apex of primary rhinarium of ultimate ant. segm. .... *Mimeuria* Börner (p. 163)
– Tarsi two-segmented. Processus terminalis projecting beyond apex of primary rhinarium of ultimate ant. segm. by more than 0.015 mm. .... *Prociphilus* subg. *Stagona* Koch (p. 159)
3 (1) Wax gland plates present on head. Apical segm. of rostrum typically with 4 or more accessory hairs. 4
– Wax gland plates absent from head, or at most one vestigial pair present as small pale spots around median occipital hairs. Apical segm. of rostrum usually without, rarely with 2–3, accessory hairs. 5
4 (3) Posterior part of abdomen abruptly narrowed and produced into an extension (Fig. 169). .... *Prociphilus* Koch s. str. (p. 154)
– Posterior part of abdomen not abruptly narrowed, nor produced, but with rounded apex. .... *Thecabius* Koch s. str. (p. 167)
5 (3) On roots of *Picea*. Ant. segm. II with 5–10 hairs. .... *Pachypappa* Koch (p. 143)
– Not on roots of *Picea*. Ant. segm. II with 2–8 hairs. 6
6 (5) Wax gland plates on abd. tergite III large, their diameter at least as long as distance between spinal and pleural plates of that segment. .... *Thecabius* subg. *Parathecabius* Börner (p. 169)
– Wax gland plates on abd. tergite III small, their diameter much shorter than distance between spinal and pleural plates of that segment. .... *Pemphigus* Hartig (p. 171)

Alate viviparous females from secondary hosts (sexuparae)
(except *Pachypappella* and *Gootiella*)

1 Antennae with secondary rhinaria on segm. V, sometimes also on segm. VI; secondary rhinaria with distinctly dotted or ciliated borders. 2

– Antennae without secondary rhinaria on segm. V and VI, or, if with 1–3 on segm. V, then rhinaria without dotted or ciliated borders. 4

2 (1) Ant. segm. VI with secondary rhinaria. Apical segment of rostrum with at least a narrow pale subapical zone (Fig. 165 A), with accessory hairs up to 0.045 mm long. *Prociphilus* Koch s. str. (p. 154)

– Ant. segm. VI without secondary rhinaria. Apical segment of rostrum without pale subapical zone (Fig. 165 B), with accessory hairs up to 0.025 mm long. 3

3 (2) Processus terminalis vestigial, not projecting more than 0.01 mm beyond the narrow crescentic primary rhinarium. Basal part of ant. segm. VI with more than 20 hairs. .... *Mimeuria* Börner (p. 163)

– Processus terminalis projecting at least 0.03 mm beyond the roundish or subquadrate primary rhinarium. Basal part of ant. segm. VI with less than 10 hairs. .... *Prociphilus* subg. *Stagona* Koch (p. 159)

4 (1) Ant. segm. III with 15–28 narrow secondary rhinaria, many of which cover half or more of the circumference of the segment (Fig. 190). Mesonotum with 2 spinal and 2 pleural wax gland plates. .... *Thecabius* Koch s. str. (p. 167)

– Ant. segm. III with 2–13 transverse oval to narrow, spindle-shaped secondary rhinaria, relatively few of which cover half or more of the circumference of the segment. Mesonotum without wax gland plates or with only two spinal plates. 5

5 (4) Secondary rhinaria of ant. segm. IV with borders more or less distinctly dotted. .... *Pachypappa* Koch (p. 143)

– Secondary rhinaria without dotted borders. 6

6 (5) First tarsal segment usually with 2–3 hairs. Ant. segm. III with 3–10 not very narrow secondary rhinaria. .... *Pemphigus* Hartig (p. 171)

– First tarsal segment often with more than 3 hairs. Ant. segm. III with 6–12 very narrow secondary rhinaria. *Thecabius* subg. *Parathecabius* Börner (p. 169)

TRIBE PROCIPHILINI

Newborn virginogeniae with long, curved empodial hairs, which are often as long as or longer than the claws (except in *Mimeuria,* see note on p. 165). Leaf nests, curled leaves, or blister-shaped or purselike galls are produced in spring on *Populus, Fraxinus, Lonicera,* and other woody angiosperms. Secondary hosts are trees, in most cases coniferous trees (or unknown). Media of fore wing with one fork or unbranched.

Genus *Pachypappa* Koch, 1856

Pachypappa Koch, 1856: 269.
Type-species: *Pachypappa marsupialis* Koch, 1856.
Asiphum Koch, 1856: 246.
Type-species: *Asiphum populi* (Fabricius) sensu Koch, 1856
= *Aphis tremulae* Linné, 1761.
Survey: 329 and 94.

Fundatrix without wax gland plates, long-haired, without siphuncular pores. Primary rhinaria surrounded by short hairs. Media of fore wing usually with one fork in fundatrigeniae, unbranched in sexuparae.

Eastop & Hille Ris Lambers (1976) and most other authors regard *Asiphum* as a separate genus, but I follow Stroyan (1975).

Nine species in the world, four species in Scandinavia. They are visited by ants on *Populus*, the primary host, in spring.

Key to species of *Pachypappa*

Fundatrices and galls
(based on information from Danielsson (in litt.))

1 Processus terminalis not distinctly pointed. Blister-shaped leaf galls not produced. 2
– Processus terminalis usually distinctly pointed. Blister-shaped leaf galls produced. 3
2 (1) Dorsum with few hairs. Frontal hairs maximally 0.075 mm. Body shorter than 4.0 mm. On *Populus alba* and *P. canescens*. 28. *warschavensis* (Nasonov)
– Dorsum with numerous hairs. Frontal hairs maximally about 0.145 mm. Body longer than 5.0 mm. On *Populus tremula*. .... 26. *tremulae* (Linné)
3 (2) Body orange with grey dorsum, usually longer than 5.0 mm. Head capsule weakly sclerotized, in mounted specimens yellowish. On *Populus tremula*. .... 25. *populi* (Linné)
– Body orange, also on dorsum, shorter than 5.0 mm. Head capsule strongly sclerotized, in mounted specimens dark brown. On *Populus alba* and *P. canescens*. .... 27. *vesicalis* Koch

Alate viviparous females from primary hosts (fundatrigeniae)
(prepared by R. Danielsson)

1 Membrane of fore wing without hairs. Genital plate with 6–12 hairs on anterior half. Abdominal spinal wax gland plates distinctly delimited. 2

– Membrane of fore wing with hairs. Genital plate with more than 10 hairs on anterior half. Abdominal spinal wax gland plates on tergites II–VIII hardly visible or absent. 3

2 (1) Apical segm. of rostrum about 0.5 × 2sht. Ant. segm. III with 5–11 secondary rhinaria on distal 2/3. .... 26. *tremulae* (Linné)

– Apical segm. of rostrum 0.7 × 2sht. or a little longer. Ant. segm. III with 4–6 secondary rhinaria on distal half. .. 28. *warschavensis* (Nasonov)

3 (1) Cauda with more than 45 hairs. Abdomen with marginal and spinal wax gland plates indistinctly delimited and difficult to observe. .... 25. *populi* (Linné)

– Cauda with 15–30 hairs. Marginal wax gland plates of abd. segm. I–VII distinctly delimited, spinal plates absent from segm. II–VIII (Fig. 156). .... 27. *vesicalis* Koch

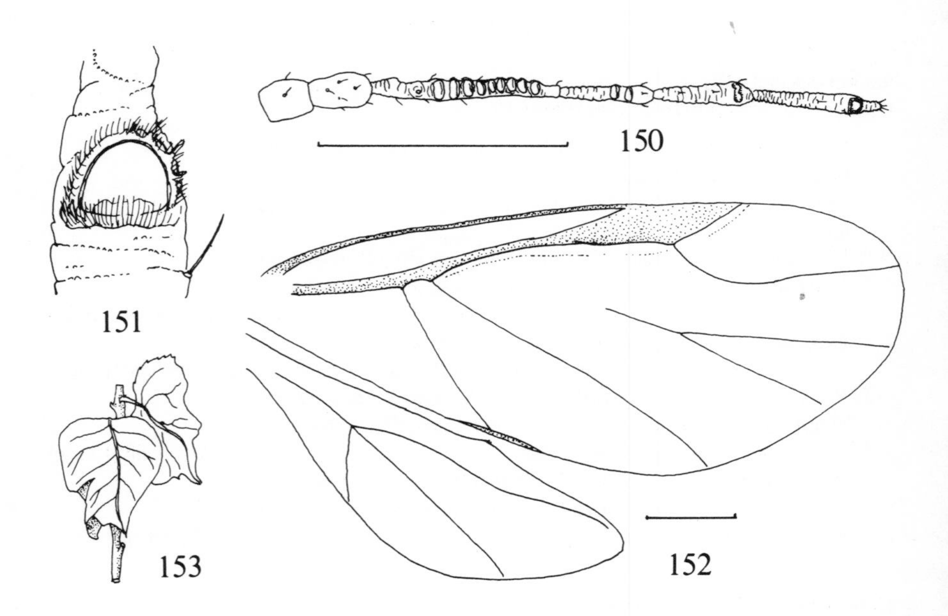

Figs. 150, 151. *Pachypappa populi* (L.), alate fundatrigenia. – 150: antenna; 151: primary rhinarium of segm. VI; 152: wings. (Scale 0.5 mm).

Fig. 153. Same, gall on *Populus tremula*. (After Danielsson, redrawn).

25. ***Pachypappa populi*** (Linné, 1758)
Figs. 150–153.

Aphis populi Linné, 1758: 453.
Pachypappa grandis Tullgren, 1925: 17.
Survey: 329.

Fundatrix. Grey. Antennae and legs brownish. Faintly wax powdered. Body almost globular, densely covered with hairs as *vesicalis*. Antenna usually 5-segmented, 1.29–1.60 × the interval between the eyes, shorter than 0.2 × body. Very similar to *vesicalis*. 5.1–6.3 mm.

Alate viviparous female from primary host (fundatrigenia). Abdomen greenish. Body with many fine hairs. Wax gland plates absent from head and thorax; six longitudinal rows of indistinct and weakly delimited plates of various size and shape present on abdomen. Antenna 6-segmented, about 0.3 × body; segm. III about as long as segm. IV+V; processus terminalis about 0.25 × VIa; primary rhinaria (Fig. 151) as in *vesicalis;* secondary rhinaria transverse oval, narrow, not surrounded by rings of short hairs, on III: 8–13, IV: 2–4. About 3.5 mm.

Distribution. Not in Denmark; in Sweden from Halland and Småland in the south, north to Vrm., Vstm., and Upl.; recorded from Norway; in Finland from Ab north to Kb. – NW Russia, Poland, the Alps. This rare species is apparently boreo-alpine. Not in N Germany or Britain.

Biology. Large, blister-shaped leaf galls, which are open on the underside, are produced in spring on *Populus tremula* (Fig. 153). The gall contains the fundatrix and its offspring; it can be 10 cm long if 2–3 fundatrices occur on the same leaf. The leaf is pale green or yellow, often a little reddish, and sack-shaped. All fundatrigeniae are alate and leave *Populus,* in Sweden in the beginning of July, usually 1–3 weeks later than the fundatrigeniae of *tremulae* (Danielsson 1976). *Picea abies* is the secondary host (Danielsson, unpublished). Occurrence of woolly virginogeniae on roots of *Picea* in spring and early summer demonstrates that overwintering can take place here.

26. ***Pachypappa tremulae*** (Linné, 1761)
Figs. 154, 157, 159.

Aphis tremulae Linné, 1761: 261.
Aphis lanata Zetterstedt, 1840: 311 (praeocc.).
Rhizomaria piceae Hartig, 1857: 52; *Pemphigus piceae* (Hartig) of Tullgren 1909: 138.
Survey: 95 (as *Asiphum tremulae;* here Stroyan (1975) is followed).

Fundatrix. Reddish brown; surface somewhat silver-like, with more or less distinct, dark, dorsal, transverse stripes, which are folds of the cuticle. Body almost globular, with flat underside, densely clothed with long fine hairs, without wax gland plates. Antenna 5-segmented, about 0.10–0.12 × body; segm. III about 1.5 × segm. II and shorter than segm. IV+V; processus terminalis about 0.10–0.11 × Va. Rostrum reaching to or just past middle coxae; apical segm. 0.8–0.9 × 2sht., without accessory hairs, with pale subapical zone. Siphuncular pores absent. About 6.0–6.5 mm.

Alate viviparous female from primary host (fundatrigenia). Abdomen orange or reddish brown, covered with wax. Body with long hairs. Wax gland plates absent from head, pairs of spinal plates present on thorax and abdomen, marginal plates also present on abdomen (Fig. 157), smaller pleural plates sometimes present on one or more segments. Antenna 6-segmented, about one third of body length; segm. III about as long as IV+V; VI slender and longer than V; V slightly longer than IV; processus terminalis 0.11–0.15 × VIa; segm. III with 5–11 transverse oval, narrow secondary

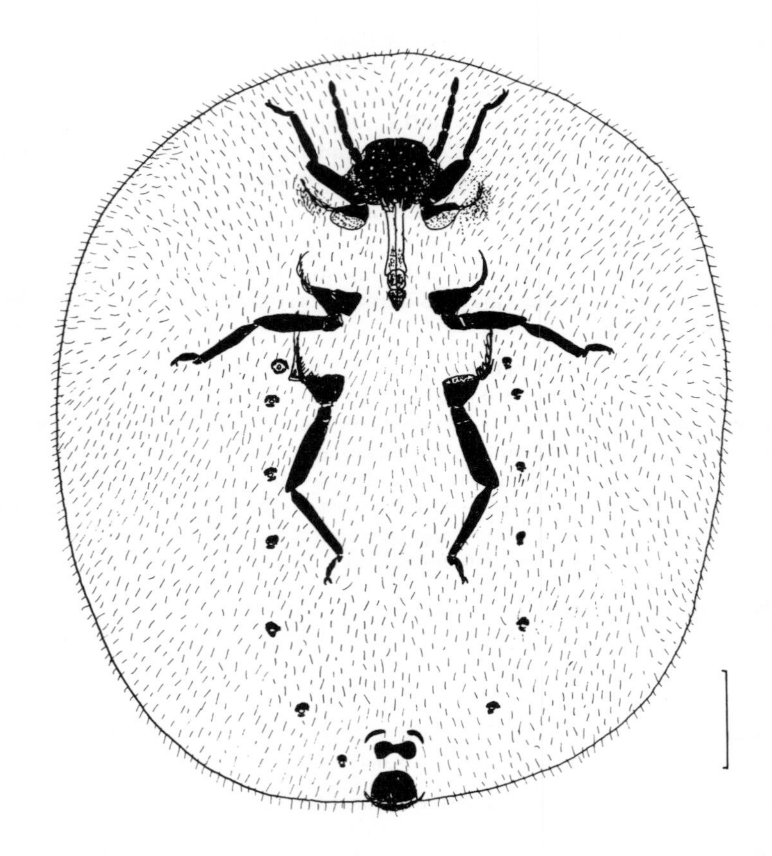

Fig. 154. Fundatrix of *Pachypappa tremulae* (L.) as seen on slide flattened and transparent so that both upper and underside are visible. (Scale 1 mm).

rhinaria, some of which have dotted borders (Fig. 159); segm. IV with 0–4, V with 0–1, VI without secondary rhinaria. Rostrum reaching middle coxae; apical segm. 0.5 × 2sht., with 0–3 accessory hairs. Siphuncular pores very small or invisible (Fig. 157, s). 3.2–4.4 mm.

Apterous viviparous female on secondary host (virginogenia). White or yellowish white; posterior part of body covered with wax. Head, antennae, and legs brown. Body with very few fine hairs. Wax gland plates absent from head and thorax, present on abdomen; each of the abd. segm. III–VI with 4 large wax gland plates, VII with 2 plates. Antenna 5-segmented, about 0.3 × body; segm. V slightly longer than III, which is twice as long as IV; II slightly longer than IV. Siphuncular pores absent. 1.3–2.1 mm.

Alate viviparous female from secondary host (sexupara). Abdomen yellowish green, with grey wax powder. Wax gland plates present on thorax and abdomen, occasionally also on posterior part of head (small). Antenna usually 6-segmented, slightly shorter than 0.5 × body; segm. III with 2–7 transverse oval secondary rhinaria with dotted borders, usually all on distal half, IV with 1–3 near apex; processus terminalis 0.16–0.24 × VIa. Apical segm. of rostrum 0.5–0.6 × 2sht., with 0(–2) accessory hairs. Siphuncular pores absent. 1.2–2.7 mm.

Distribution. Only few records from Denmark, where it has been found in SJ, NWJ, and NEJ; in Sweden common from Sk. north to T. Lpm.; in Norway known from AK, Bø, O, and HE; in Finland known from Al, Ab, and N north to Li, in some years the galls are very common (Heikinheimo in litt.). – Widespread in Europe, from Great Britain east to the forest zone of Russia; obviously rare in N Germany (F. P. Müller in litt.); in Poland also at the Baltic coast; W Siberia; N America.

Biology. In spring the fundatrices are found on the bark of short twigs of *Populus tremula.* The offspring move to the new shoots and cause curving of the petioles, whereby the leaves are bent towards each others to form a nest, but they are not deformed. Ants have been observed visiting the leaf nests. All the nymphs become alate fundatrigeniae which fly to *Picea abies,* the secondary host. Their offspring are born above the soil and move to the roots, where several generations of apterous viviparous females feed, and are visited by ants. The alate sexuparae are born on the *Picea* roots and return to aspen in September-October. Overwintering of apterae on roots of *Picea* may take place. Tullgren (1909) observed root aphids as early as in the middle of May in Östergötland. I have seen them in April in Denmark.

27. ***Pachypappa vesicalis*** Koch, 1856
Figs. 155, 156.

Pachypappa vesicalis Koch, 1856: 272.
Survey: 330.

Fundatrix. Reddish yellow. Head, antennae, and legs brown. Covered with a thin layer of white wax. Body almost globular, often broader than long, densely haired; hairs on head and thorax relatively long. Wax gland plates absent. Antenna 5-segmented,

1.19–1.26 × the interval between eyes; segm. V about twice as long as IV. Rostrum reaching past middle coxae. Siphuncular pores absent. 4.2–4.9 mm.

Alate viviparous female from primary host (fundatrigenia). Abdomen reddish yellow, wax powdered. Body with many fine hairs. Wax gland plates absent from head and pro- and mesothorax; marginal wax gland plates consisting of pentagonal facets present on abdomen (Figs. 155, 156). Antenna 6-segmented, about 0.3 × body; segm. III about as long as IV+V; processus terminalis about 0.3 × VIa; primary rhinaria very small, surrounded by short hairs; secondary rhinaria transverse oval, not surrounded by short hairs, on III: 10–13, IV: about 3, V and VI: 0. Siphuncular pores present. About 3.3 mm.

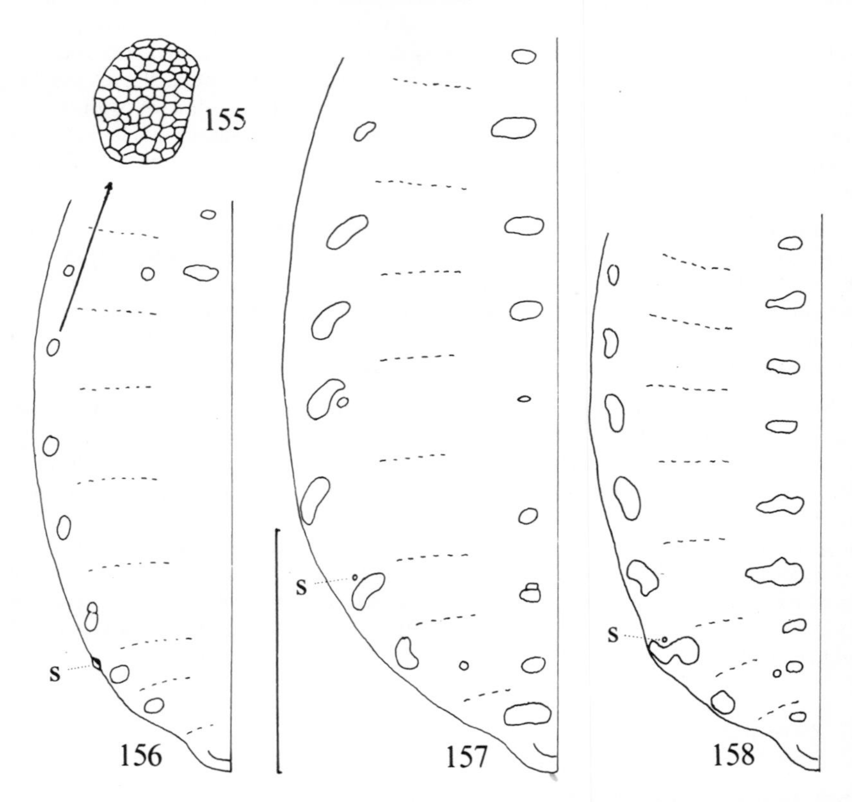

Figs. 155–158. Wax gland plates of alate fundatrigeniae of *Pachypappa* spp. – 155: *vesicalis* Koch, marginal plate on abd. segm. II; 156–158: arrangement of plates on one side of metathorax and abdomen of 156: *vesicalis* Koch; 157: *tremulae* (L.); 158: *warschavensis* (Nas.); schematical, dorsal view. – s = siphuncular pore. (Scale 0.1 mm for 155, 1 mm for the rest).

Distribution. In Sweden recorded from Sk. north to Upl., including Öl.; not in Denmark, Norway, or Finland. – Known from Germany, including northern DDR and Hamburg area, but rare; also Poland (incl. Baltic coast) and NW Russia; not in Britain.

Biology. The species produces irregularly blistered, sack-shaped leaf galls on *Populus alba* (and *P. canescens*). Three or four galls often occur close together in the top of the shoot. The gall is first green and gradually becomes yellowish. The fundatrigeniae are alate and leave the primary host, in Sweden even in June (Danielsson 1976). They fly to the secondary host, which is *Picea (abies, glauca)*. The alate sexupara, which returns to *Populus* in autumn, is reported to have only 2 rhinaria on ant. segm. III and IV (Börner & Heinze 1957).

28. ***Pachypappa warschavensis*** (Nasonov, 1894)
Fig. 158.

Pemphigus warschavensis Nasonov, 1894: 17.
Pemphigus varsoviensis Mordvilko, 1895: 183.
Survey: 330 (as *Pachypappa varsoviensis*).
Fundatrix. Head and body with few hairs, which are shorter than in *tremulae;* those on frons maximally 0.075 mm long. Otherwise similar to *tremulae,* but smaller, body length less than 4.0 mm.

Alate viviparous female from primary host (fundatrigenia) (Fig. 158). Very similar to *tremulae,* but media with very short fork, frequently unbranched. Apical segm. of rostrum 0.7–0.75 × 2sht. Secondary rhinaria placed distally on ant. segm., III: 4–6 (usually 5), IV: 1–2, V: 0–2, VI: 0. 3.0–3.1 mm.

Distribution. Recently found in Sk. in Sweden by Danielsson (unpublished); not in Denmark, Norway, or Finland. – Recorded from Poland, Germany, Spain, Transcaucasia, and W Siberia, apparently very rare.

Biology. The primary hosts are *Populus alba* and – according to Shaposhnikov (1964) – also *P. canescens* and *P. hybrida.* Leaf nests similar to those of *tremulae* are formed (Börner & Heinze 1957).

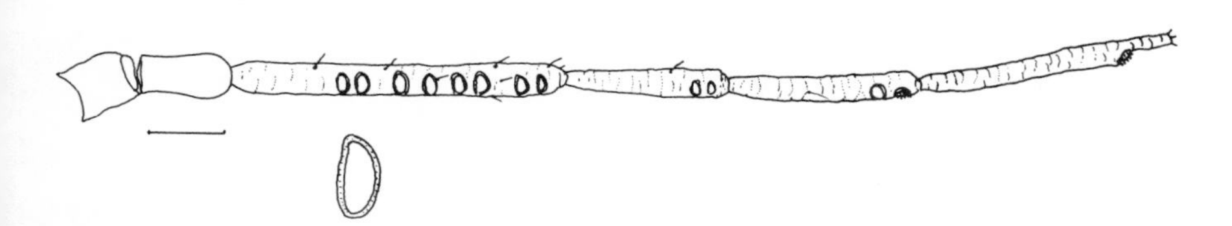

Fig. 159. Antenna of alate fundatrigenia of *Pachypappa tremulae* (L.). (Scale 0.1 mm). A secondary rhinarium with dotted border is shown enlarged below.

Note. The synonymy was pointed out by Szelegiewicz (1978). The only material examined by me are three fundatrigeniae collected in Sweden by Danielsson, who prepares a paper on this little-known species. The species has not been considered in the key to genera. The wing veins of the specimens studied are paler than in other *Pachypappa* spp., and the media is unbranched. These specimens will key out as *Mimeuria*. There are fewer secondary rhinaria than in *Mimeuria*, and the apical segm. of rostrum is relatively slightly longer (0.5–0.7 × 2sht. in *Mimeuria*) with a lower average number of accessory hairs (about 2; *Mimeuria:* about 4).

Genus *Pachypappella* Baker, 1920

Pachypappella Baker, 1920: 71.
Type-species: *Pachypappa lactea* Tullgren, 1909.
Survey: 330.

Fundatrix with wax gland plates on thorax and abdomen, but not on head, short-haired, otherwise similar to *Pachypappa*.

Two species in the world, one species in Scandinavia.

29. ***Pachypappella lactea*** (Tullgren, 1909)
Figs. 160, 161.

Pachypappa lactea Tullgren, 1909: 69. – Survey: 330.

Fundatrix. Greyish brown (in alcohol yellowish orange), covered with snow-white wax, carrying a large tuft of wax on the posterior part of the body; head and legs blackish brown. Body with rather few short hairs except for on head and apex of abdomen.

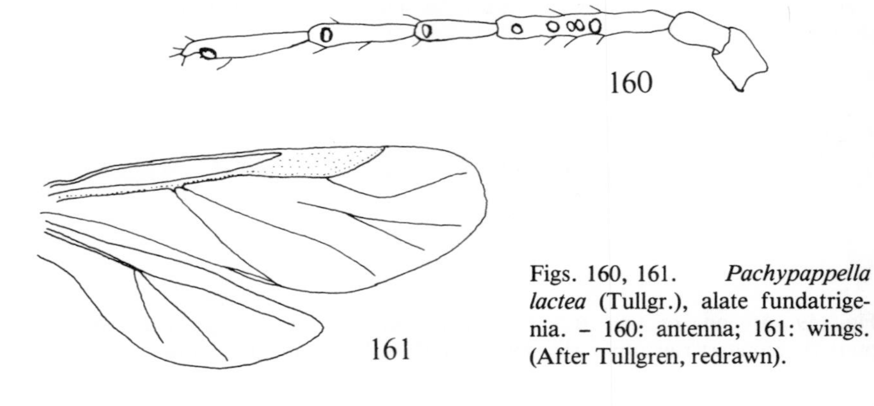

Figs. 160, 161. *Pachypappella lactea* (Tullgr.), alate fundatrigenia. – 160: antenna; 161: wings. (After Tullgren, redrawn).

Marginal wax gland plates, which are circular or oval and distinctly delimited, are present on thorax and abd. segm. I–VII. Antenna 5-segmented, about 0.2 × body; segm. III about as long as IV+V; V at least twice as long as IV; processus terminalis about 0.2–0.3 × Va; primary rhinaria small, surrounded by short hairs. Rostrum very short. About 4 mm.

Alate viviparous female from primary host (fundatrigenia). Abdomen dark greyish-brown. Body hairs rather long. Mesonotum with 2 wax gland plates; abd. tergites I–VII usually each with 2 pairs of plates, but spinal plates rather small; abd. tergite VIII with 2 large plates. Antenna 6-segmented, about 0.4 × body; segm. III about as long as IV+V; V slightly longer than IV and a little shorter than VI; III with a small toothlike process near base; processus terminalis about 0.2 × VIa; primary rhinaria as in fundatrix; III with 5–10 oval, rather broad secondary rhinaria, IV with 1–2, V and VI without secondary rhinaria. Wing veins brownish. Siphuncular pores rather large. 2.5–3.5 mm.

Distribution. Not in Denmark; in Sweden occurring from Sk. north to Nb., in some years found in large numbers on the primary host (Danielsson 1976, Lampel 1970); not in Norway; all over the southern part of Finland, also in Li in the north, and the galls very common in some years (Nuorteva 1961 and Heikinheimo in litt.). – Recorded from the Alps (Austria), the Baltic region of Russia, and N Asia (W Siberia, Mongolia); not in Britain, N Germany, or Poland. The species seems to be boreo-alpine.

Biology. The primary host is *Populus tremula.* Fundatrix and its offspring are situated inside a leaf gall, which is often placed at the base of a shoot. The leaf is red or yellow and folded along the midrib towards the underside, and the petiole is often sharply bent. Aphids may occur also on the upper side of the leaf, outside the gall, when colonies become very large (Danielsson 1976). The other leaves of the shoot eventually turn yellow or orange. Most alate fundatrigeniae leave the aspen in late June or early July in Sweden (Danielsson 1976) and fly to *Picea abies* (Danielsson, unpublished), but some can still be found in the neighbourhood of the galls on *Populus* in August (Tullgren 1909).

Genus *Gootiella* Tullgren, 1925

Gootiella Tullgren, 1925: 22.

Type-species: *Gootiella tremulae* Tullgren, 1925.

Survey: 208.

Fundatrix with wax gland plates also on head in addition to those on thorax and abdomen. Primary rhinaria not surrounded by distinct rings of short hairs, very large in the alate fundatrigenia.

Two species in the world, one species in Scandinavia. The other species, *G. alba* Shaposhnikov, lives on *Populus alba* in the USSR and is little known (no alate morphs described).

30. ***Gootiella tremulae*** Tullgren, 1925
Figs. 162–164.

Gootiella tremulae Tullgren, 1925: 24. – Survey: 208.

Fundatrix. Reddish yellow, covered with wax. Body almost globular, nearly hairless; a few short, fine hairs present on head, apex of abdomen, antennae, and legs. Circular or oval, distinctly delimited wax gland plates present on head: 4 small plates on the anterior part, 2 large ones on the posterior part; on prothorax: 1 spinal and 1 marginal pair of plates; on meso- and metathorax and abd. segm. I–VI: 1 spinal, 1 pleural, and 1 marginal pair per segment; on abd. segm. VII: 4 plates. Antenna 5-segmented, shorter than 0.2 × body; segm. V about 1.5 × IV. Rostrum reaching past middle coxae.

Alate viviparous female from primary host (fundatrigenia). Abdomen reddish yellow. Body with rather few short hairs. Wax gland plates absent from head and thorax, present on abdomen; abd. segm. I–V with pleural and marginal rows of plates, each of which is subdivided into several small, facetted plates (Fig. 162); the pleural plates may be absent from some segments. Antenna 6-segmented, about 0.3 × body; segm. III as long as or a little longer than IV+V; VI longer than V, thickest in the middle; primary rhinaria very large, star-shaped (Fig. 163); III with 12–15 rather large, transverse oval secondary rhinaria, IV with 3–4, V and VI without secondary rhinaria. Apical segm. of rostrum about 0.5 × 2sht. Siphuncular pores present, but rather indistinct. About 3.3–3.9 mm.

Distribution. In Denmark found only once, viz. at Lyngby (NEZ) in August 1940; Sweden: Tullgren's record from Sörby in Östergötland (1912) was the only finding until

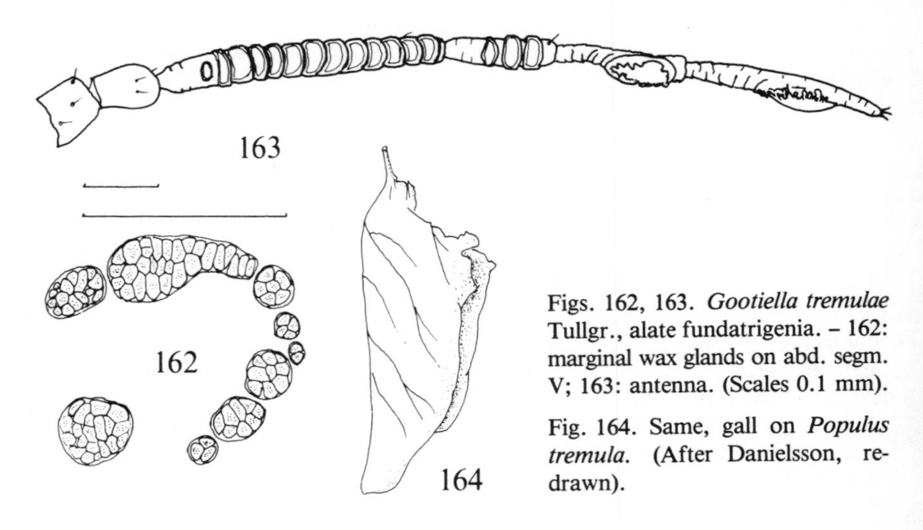

Figs. 162, 163. *Gootiella tremulae* Tullgr., alate fundatrigenia. – 162: marginal wax glands on abd. segm. V; 163: antenna. (Scales 0.1 mm).

Fig. 164. Same, gall on *Populus tremula*. (After Danielsson, redrawn).

recently, when Danielsson (in litt.) added records from Sk., Sm., and Öl.; the species is also known from Norway and Finland (N). – Outside Scandinavia only recorded from Poland (rare).

Biology. The primary host is *Populus tremula.* An extremely large, more than 10 cm long, sack- or purse-shaped leaf gall (Fig. 164) is produced, usually in the top of high trees. The leaf is folded so that its two halves are closed together. First the gall is pale green, later it becomes yellowish. Fundatrix and fundatrigeniae develop inside the gall. Alate fundatrigeniae have been observed in August in Denmark. The secondary host is *Juniperus communis* (Danielsson 1976). Overwintering of virginogeniae on roots of *Juniperus* is probably commoner than overwintering of eggs on *Populus.*

Genus *Prociphilus* Koch, 1857 s. lat.

Prociphilus Koch, 1857: 279.
Type-species: *Aphis bumeliae* Schrank, 1801.
Survey: 358.

Wax gland plates present on head, thorax, and abdomen, also in fundatrix. Wax production considerable. Media of fore wing unbranched. Primary rhinaria surrounded by short hairs. Tarsi and tarsal claws relatively long. Siphuncular pores small or absent.

The primary hosts are not species of *Populus* as in the other genera of Pemphiginae, but members of Oleaceae, Caprifoliaceae, Aceraceae, and Pomaceae. The secondary hosts of Scandinavian species are coniferous trees as in other genera of Prociphilini. The virginogeniae feed on the roots of the secondary hosts and can hibernate there (anholocyclically). The alate fundatrigeniae, which fly from the primary to the secondary hosts, bear their young (as "pseudoeggs" (Hille Ris Lambers 1950b) which "hatch" shortly after) on needles during a brief period. These young have a relatively long rostrum, an adaptation to feeding in bark crevices.

The genus contains 45 species, but only four of these occur in Scandinavia, belonging to two subgenera. They are not visited by ants.

Key to subgenera of *Prociphilus*

All morphs except sexuales

1 Apical segm. of rostrum with at least a narrow pale subapical zone (Fig. 165 A). .... *Prociphilus* Koch s. str. (p. 154)
– Apical segm. of rostrum without pale subapical zone (Fig. 165 B). .... *Stagona* Koch (p. 159)

Subgenus *Prociphilus* Koch, 1857 s. str.

Apical segm. of rostrum with at least a narrow pale subapical zone (Fig. 165 A). Pterostigma of normal appearance (Fig. 172 A). Sexuparae with secondary rhinaria on ant. segm. VI (Fig. 168).

With 33 species in the world, 2 species in Scandinavia, both host-alternating between *Fraxinus* (I) and *Abies* (II).

Key to species of *Prociphilus* s. str.

Alate viviparous females (fundatrigeniae and sexuparae)

1 Ant. segm. III 6–7 × segm. II, about as long as IV+V; IV, V, and VI about equally long (Figs. 166 and 168). Wax gland plates between eyes better developed than anterior plates between antennae, which are often absent (Fig. 171 A). .... 31. *bumeliae* (Schrank)

– Ant. segm. III 3–5 × segm. II, shorter than IV+V; V shorter than VI, slightly longer than IV (Fig. 167). Wax gland plates on head between antennae better developed than posterior plates between eyes, which are often absent (Fig. 171 B). .... 32. *fraxini* (Fabricius)

Apterous viviparous females on secondary hosts

1 Posterior pair of wax gland plates on head better developed than anterior pair. .... 31. *bumeliae* (Schrank)

– Posterior pair of wax gland plates on head not better developed than anterior plates, sometimes absent. .... 32. *fraxini* (Fabricius)

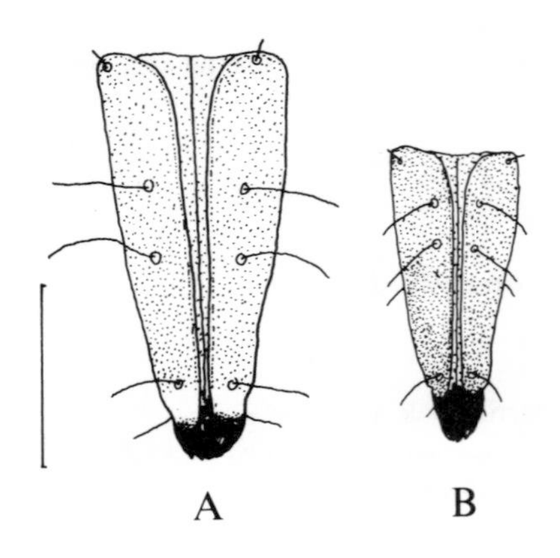

Fig. 165. Apical segment of rostrum of fundatrices. – A: *Prociphilus bumeliae* (Schr.), with pale subapical zone; B: *P. (Stagona) xylostei* (DeGeer), without pale subapical zone. (Scale 0.1 mm).

31. ***Prociphilus (Prociphilus) bumeliae*** (Schrank, 1801)
Figs. 165 A, 166, 168–170, 172 A.

Aphis bumeliae Schrank, 1801: 102.
Pemphigus poschingeri Holzner, 1874: 221.
Survey: 359.

Fundatrix. Brown, covered with white wax wool. Body with only few very fine hairs. Wax gland plates on head and prothorax very distinct, on abdomen large, but indistinct; on head 2 large plates between the eyes, in front of them several smaller plates; on prothorax 4 large plates in one transverse row. Antenna 5-segmented, about 0.3 × body; segm. III longer than IV+V; V longer than IV; processus terminalis about 0.2 × Va. Rostrum reaches to or a little past middle coxae. Legs rather thick. About 4 mm.

Alate viviparous female from primary host (fundatrigenia). Head and thorax blackish brown, with white spots corresponding to the wax gland plates, abdomen light brown or yellowish red. Head with only the posterior pair of wax gland plates well developed (Fig. 171 A); two very small plates sometimes present more anteriorly between the antennae. Antenna (Fig. 166) 6-segmented, about 0.5 × body; segm. III 6–7 × segm. II, about as long as IV+V; IV, V, and VI about equally long; processus terminalis about 0.15 × VIa; III with 16–24 short, transverse oval secondary rhinaria, IV with 4–9, V and VI without secondary rhinaria. Apical segm. of rostrum about 0.6 × 2sht., with about 6 accessory hairs. First tarsal segm. with (3 or) 4 hairs. Siphuncular pores absent. 3.8–5.5 mm.

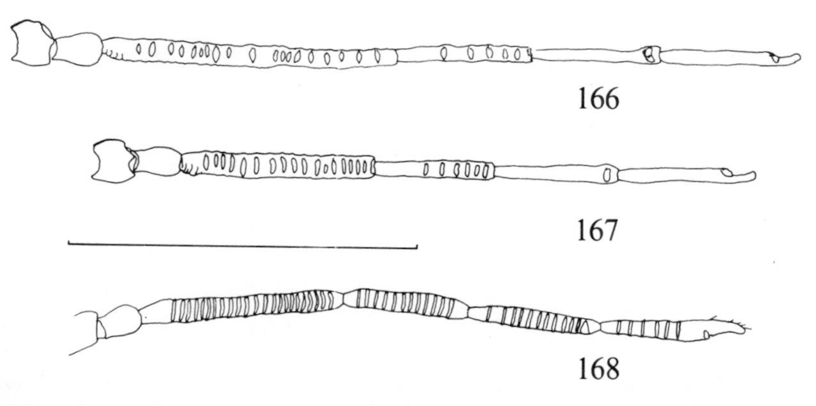

Figs. 166–168. Antennae of alate viviparous females of *Prociphilus* spp. – 166: *bumeliae* (Schr.), fundatrigenia; 167: *fraxini* (F.), fundatrigenia; 168: *bumeliae*, sexupara. (168 after Tullgren, redrawn). (Scale 1 mm).

Apterous viviparous female on secondary host (virginogenia). Covered with wax. Wax gland plates present on head, thorax, and abdomen; on head only the posterior pair of plates distinctly developed; the anterior plates may be absent. Body almost hairless. Antenna 6-segmented, about 0.25 × body; segm. VI longer than V, which is slightly longer than III; processus terminalis about 0.2 × VIa. Rostrum reaching past middle coxae; apical segm. 1.1 × 2sht. Posterior part of abdomen (with cauda and anal plate) narrow, cauda-like, about 1.5 times as long as wide (Fig. 169). About 2.9 mm. – First instar nymph of virginogenia born by fundatrigenia (Fig. 170): Body and appendages with numerous long hairs. Antenna 4-segmented. Rostrum extremely long, about 1.5 × body length. About 1.0 mm. – Nymphs of later generations of virginogeniae have very short hairs.

Alate viviparous female from secondary host (sexupara). Wax gland plates as in fundatrigenia. Antenna (Fig. 168) 0.6–0.7 × body; segm. III of about the same length as IV+V or slightly shorter; secondary rhinaria: III: 23–30, IV: 9–14, V: 10–16, VI: 5–9; all rhinaria surrounded by very short hairs. 2.1–3.1 mm.

Distribution. In Denmark found in NWJ and NEJ, but probably occurring all over the country; in Sweden from Sk. north to Upl.; not in Norway and Finland. – Widespread in W, C & E Europe, e.g. Britain, Germany (not N Germany), Poland, NW & W Russia, Switzerland, Czechoslovakia, Hungary, and Bulgaria.

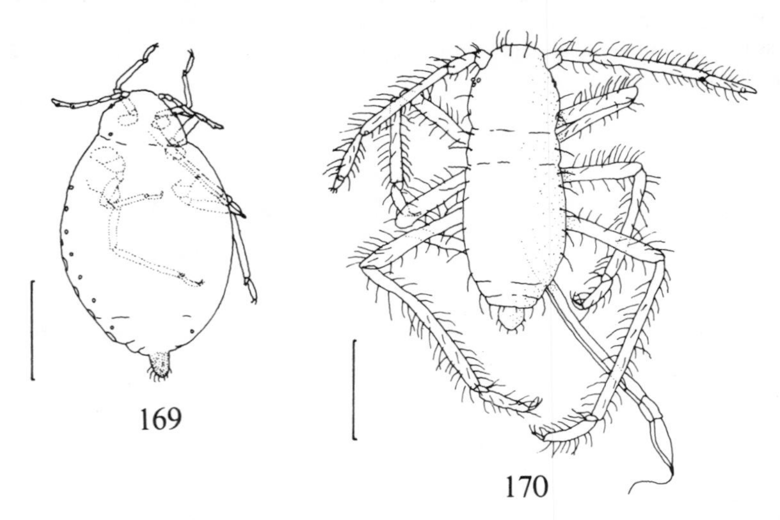

Fig. 169. *Prociphilus bumeliae* (Schr.), adult apterous female from root of *Abies* (dorsal wax gland plates omitted). (After Mordvilko, redrawn). (Scale 1 mm).

Fig. 170. Same, newborn nymph of the first generation on *Abies* (born by an alate fundatrigenia). (After Mordvilko, redrawn). (Scale 0.3 mm).

Biology. The primary host is *Fraxinus excelsior.* Some other members of Oleaceae, viz. *Syringa* and *Ligustrum* have been reported as primary hosts. Leaf nests are formed in spring. They are not as dense as those formed by *fraxini.* The fundatrigeniae fly to *Abies* (in Denmark in the end of June) and bear rather large young. The rostrum of these is considerable longer than the body. The virginogeniae feed on roots of *Abies,* and the sexuparae, which in autumn return to *Fraxinus,* are born there.

32. ***Prociphilus (Prociphilus) fraxini*** (Fabricius, 1777)
Figs. 167, 171 B.

Aphis fraxini Fabricius, 1777: 386.
Pemphigus fraxini Hartig, 1841: 367.
Prociphilus nidificus Löw, 1882: 14.
Survey: 359.

Fundatrix. Very similar to *bumeliae.*

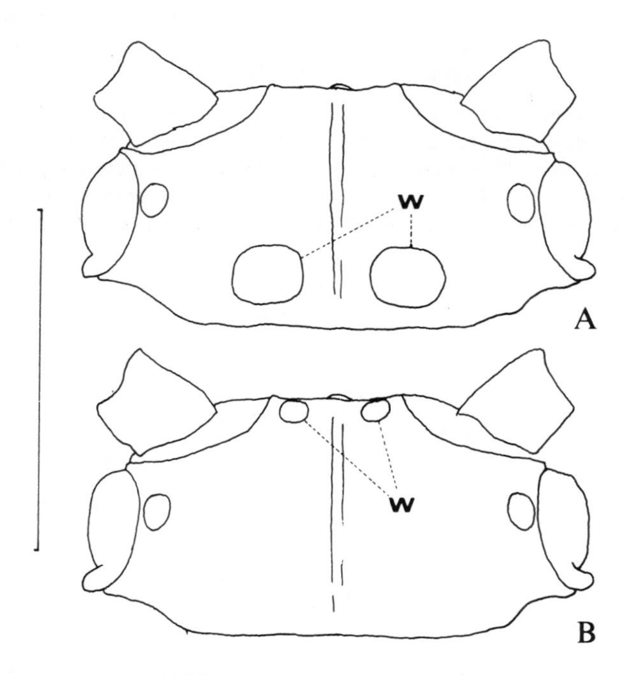

Fig. 171. Head of alate fundatrigeniae of *Prociphilus* spp. – A: *bumeliae* (Schr.); B: *fraxini* (F.); w = wax gland plates. (Scale 0.5 mm).

Alate viviparous female from primary host (fundatrigenia). Similar to *bumeliae,* but head usually with only the anterior pair of small wax gland plates between the antennae well developed; the posterior plates indistinct or absent (Fig. 171 B). Ant. segm. III shorter than in *bumeliae,* 3–5 × II, as long as or shorter than IV+V; V as long as or a little longer than IV and a little shorter than VI (Fig. 167); processus terminalis about 0.15–0.2 × VIa; secondary rhinaria: III: 14–24, IV: 1–10, V: 0–1(–4 according to Tullgren 1909), VI: 0. Apical segm. of rostrum about 0.5 × 2sht., with about 6 accessory hairs. First tarsal segm. usually with 2 or 3 hairs. Siphuncular pores absent. 3.3–5.1 mm.

Apterous viviparous female on secondary host (virginogenia). Very similar to *bumeliae,* but head with two pairs of wax gland plates, or the posterior pair absent. Rostrum reaching past middle coxae, may extend to apex of abdomen. About 1.8–2.7 mm. – First instar nymph of virginogenia born by fundatrigenia: similar to *bumeliae,* but rostrum somewhat shorter, about as long as body. 0.8–0.9 mm.

Alate viviparous female from secondary host (sexupara). Very similar to fundatrigenia, but with secondary rhinaria also on ant. segm. V and VI.

Distribution. In Denmark found in NWJ and NEZ, but probably common all over the country; in Sweden recorded from Sk. and Upl.; known from Norway; not in Finland. – Widespread in Europe from Great Britain to Poland and Russia (incl. the NW region and the Caucasus region), south to Bulgaria and Turkey.

Biology. The species is host-alternating between *Fraxinus excelsior* (I) and roots of *Abies* (II). Fundatrix is located at the base of the new shoots. Only one generation of fundatrigeniae is born. They feed as nymphs on the young shoots and petioles, which become deformed so that dense leaf nests are formed. These are usually placed high in the trees. The fundatrigeniae become alate in early summer and fly to *Abies* to bear

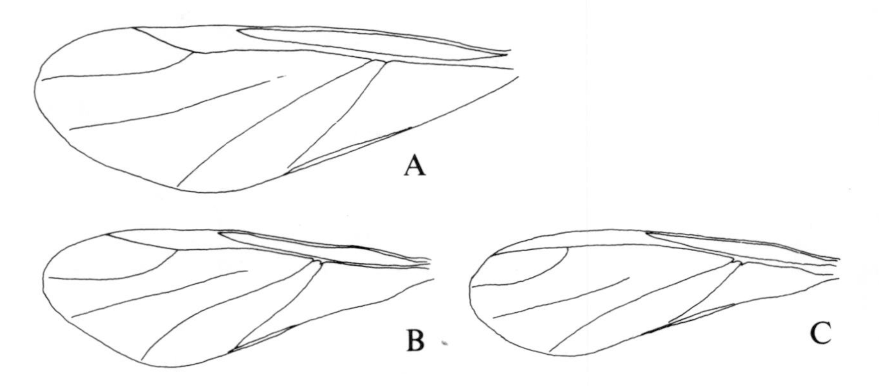

Fig. 172. Fore wings of *Prociphilus* spp. – A: *bumeliae* (Schr.), fundatrigenia; B: *xylostei* (DeGeer), fundatrigenia, and C: same, sexupara. (After Tullgren, redrawn).

longbeaked youngs, which move to the roots, where several generations of apterous virginogeniae develop. The alate sexuparae are born on the roots of *Abies* and fly to *Fraxinus* in autumn.

Subgenus *Stagona* Koch, 1857

Stagona Koch, 1857: 284.
Type-species: *Aphis xylostei* DeGeer, 1773.
Survey: 409.

Apical segm. of rostrum without a pale subapical zone (Fig. 165 B). Pterostigma somewhat prolonged, tapering (Fig. 172 B–C). Alatae, not only fundatrigeniae, but also sexuparae, without secondary rhinaria on ant. segm. VI.

With 8 species in the world, 2 species in Scandinavia.

Key to species of *Stagona* s. str.

Fundatrices

1 Ant. segm. III shorter than 1.5 × V; V longer than 1.5 × IV. On *Crataegus*. .... 33. *pini* (Burmeister)
– Ant. segm. III longer than 1.5 × V; V shorter than 1.5 × IV. On *Lonicera*. .... 34. *xylostei* (DeGeer)

Alate viviparous females

1 Head with two wax gland plates on the anterior part, sometimes also with two posterior plates (Fig. 173). Fundatrigenia with 3–11 secondary rhinaria on ant. segm. V (Fig. 174). ...... 33. *pini* (Burmeister)
– Head with two wax gland plates on the posterior part, usually without anterior plates. Fundatrigenia without secondary rhinaria on ant. segm. V (Fig. 179). .... 34. *xylostei* (DeGeer)

Apterous viviparous females on secondary hosts

1 Apical segm. of rostrum about 1.0 × 2sht. On roots of *Pinus*. .... 33. *pini* (Burmeister)
– Apical segm. of rostrum about 0.7 × 2sht. On roots of *Picea*. 34. *xylostei* (DeGeer)

33. ***Prociphilus (Stagona) pini*** (Burmeister, 1835)
Figs. 173–175.

Rhizobius pini Burmeister, 1835: 87.
Prociphilus crataegi Tullgren, 1909: 96.
Survey: 360.

Fundatrix. Green or brownish green, wax powdered. Similar to *xylostei.* Head, pronotum, antennae, and legs blackish brown. Body with rather few short, fine hairs. Wax gland plates almost as in *xylostei,* but spinal plates on metathorax and spinal and pleural plates on abdomen invisible in the single specimen available. Antenna 5-segmented, 0.25–0.3 × body; segm. III always shorter than IV+V. Rostrum reaching past middle coxae; apical segm. about 0.8 × 2sht., with 2 accessory hairs. 2.7–3.6 mm.

Alate viviparous female from primary host (fundatrigenia). Abdomen light green or greyish green. Head with 2 anterior wax gland plates, and sometimes also 2 posterior plates (Fig. 173). Wax gland plates on thorax and abdomen as in *xylostei,* but pleural and spinal plates apparently absent from abdomen. Antenna (Fig. 174) 6-segmented, about 0.5 × body; segm. III 3.8–4.2 × II, about as long as IV+V; V longer than IV, a little shorter than VI; processus terminalis about 0.25 × VIa; primary rhinarium on V rather large, transverse oval, irregular in outline; secondary rhinaria transverse oval, narrow, often rather short, not placed in one row, not parallel to each others; on III: 19–32, IV: 8–12, V: 3–11, VI: 0. Apical segm. of rostrum about 0.5 × 2sht., with 2–6 accessory hairs. First tarsal segm. usually with 2 hairs. Siphuncular pores absent. 2–3 mm.

Apterous viviparous female on secondary host (virginogenia). Wax gland plates not very distinct, present on head (maybe also on thorax and abdomen, not easy to observe in the present material). Body hairs short. Antenna 5-segmented, 0.25 × body; segm. III about as long as or a little longer than II, longer than IV and a little shorter than V; processus terminalis 0.25 × Va. Rostrum reaching past hind coxae; apical segm. about 1.0 × 2sht. 1.7–1.8 mm.

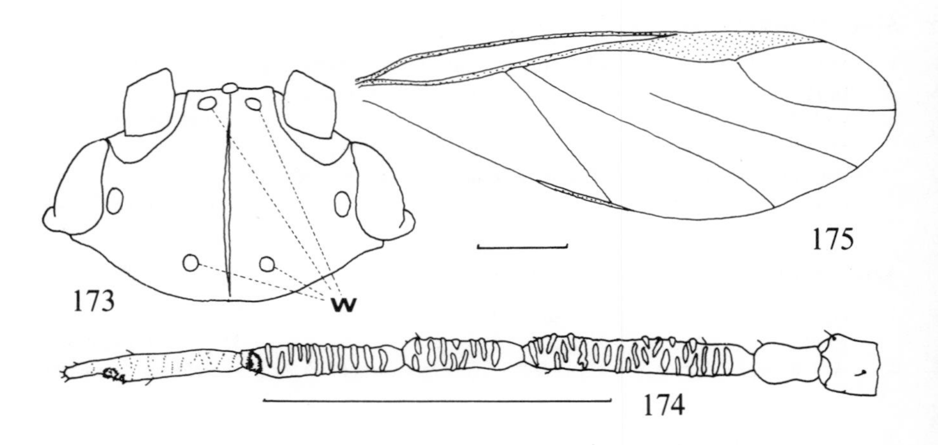

Figs. 173–175. *Prociphilus (Stagona) pini* (Burm.), alate fundatrigenia. – 173: head with wax gland plates (w) (the other three circles of about the same shape are the ocelli); 174: antenna; 175: fore wing. (Scales 0.5 mm).

Alate viviparous female from secondary host (sexupara). Very similar to fundatrigenia. Antennae longer than 0.5 × body, Secondary rhinaria: III: 16–25, IV: 9–15, V: 7–13, VI: 0. 1.1–1.7 mm.

Sexuales. Much like sexuales of *xylostei.* Oviparous female about 1 mm. Male about 0.6–0.7 mm.

Distribution. In Denmark rare, known from NWJ, NEJ, and NEZ; in Sweden widespread, from Sk. north to Vrm.; in Norway recorded from AK; in Finland recorded from N, Sa, and Sb. – Great Britain, the Netherlands, France, Switzerland, Germany (not in N Germany), Poland (uncommon), and NW Russia.

Biology. The primary host is *Crataegus.* The leaves are curled and become yellowish along the ribs. In Denmark the fundatrix can still be found in July, the fundatrigeniae in July and also as late as in August on *Crataegus.* The virginogeniae feed on roots of *Pinus* (*P. silvestris* a.o.). Sexuales have been found on *Crataegus* in Sweden in October.

34. ***Prociphilus (Stagona) xylostei*** (DeGeer, 1773)
Plate 2: 1. Figs. 165 B, 172 B–C, 176–180.

Aphis xylostei DeGeer, 1773: 96. – Survey: 361.

Fundatrix. Yellowish green, wax-powdered. Head, pronotum, antennae, and legs brown. Body with rather few short, fine hairs. Head with 2 pairs of wax gland plates (Fig. 176), the anterior plates often fused; prothorax with 2 spinal (often fused) and 2 marginal plates; meso- and metathorax and abd. segm. I–VI each with 2 spinal, 2 pleural, and 2 marginal plates; VII with 4 plates. Antenna (Fig. 178) 5-segmented, about 0.3 × body; segm. III longer than V, sometimes longer than IV+V; processus terminalis about 0.2 × Va; III with 0–2 secondary rhinaria. Apical segm. of rostrum (Fig. 165 B) about 0.7 × 2sht., with 6–11 accessory hairs. Siphuncular pores absent. 3.0–4.6 mm.

Alate viviparous female from primary host (fundatrigenia). Abdomen pale green. With 2 wax gland plates, larger than the ocelli, on posterior part of head; dwarfish specimens (the "forma *minima*" of Tullgren 1909) produced under bad nutritional conditions with additional anterior plates between the antennae (the following part of the description does not cover "forma *minima*"); thoracic segments each with (2–)4 plates; abd. segm. I–VI each with 6 plates; VII with 4 plates, VIII with 2 plates. Antenna 6-segmented, about 0.3–0.5 × body; segm. III about 4–4.5 × II, longer than IV+V; V slightly longer than IV and slightly shorter than VI; processus terminalis about 0.2 × VIa; III with 18–27 secondary rhinaria, IV with 4–8, V and VI without secondary rhinaria. Apical segm. of rostrum slender, about 0.5 × 2sht., with 4–6 accessory hairs. First tarsal segm. usually with 2 hairs. Siphuncular pores indistinct or absent. 2.3–3.7 mm.

Apterous viviparous female on secondary host (virginogenia). Wax gland plates only present on the posterior part of the abdomen (Fig. 180), placed in 4 longitudinal rows. Antenna usually 6-segmented, but segm. III and IV may be fused, about 0.5 × body. Apical segm. of rostrum about 0.7 × 2sht.

Alate viviparous female from secondary host (sexupara). Spinal and pleural wax gland plates usually absent. Ant. segm. III about 3.5–4 × II, shorter than IV+V; III with 7–22 secondary rhinaria, IV with 4–12, V with 4–13, VI without secondary rhinaria. Otherwise much like the fundatrigenia. 1.7–2.1 mm.

Sexuales. Reddish yellow. Wax glands absent. Antenna 5-segmented. Oviparous female 0.8 mm. Male 0.6 mm.

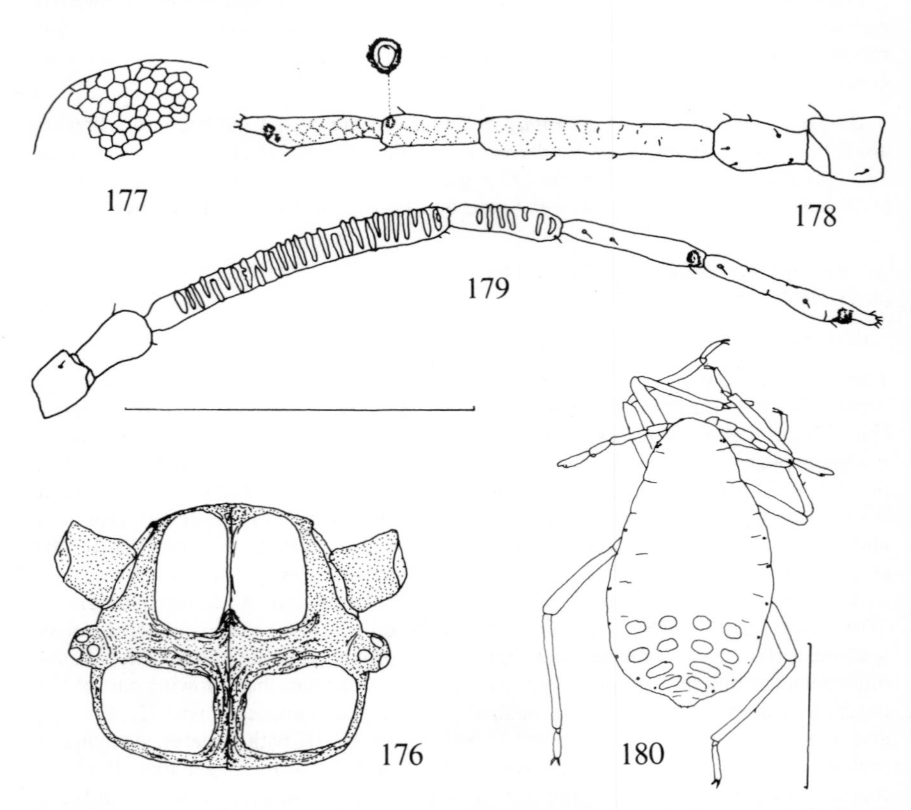

Figs. 176–178. Fundatrix of *Prociphilus (Stagona) xylostei* (DeGeer). – 176: head with wax gland plates; 177: part of abdominal wax gland plate; 178: antenna, with primary rhinarium of segm. IV on larger scale above. (Scale 0.5 mm for 176 and 178).

Fig. 179. Same, antenna of alate fundatrigenia. (Scale 0.5 mm).

Fig. 180. Same, apt. viv. from root of *Picea*. (After Mordvilko, redrawn). (Scale 1 mm).

Distribution. In Denmark rather common and widespread; in Sweden common, from Sk. north to Vb.; in Norway recorded from NTi; in Finland from the southern coast north to Ok. – Widespread in Europe, e.g. in Great Britain, Germany, the Alps, Poland, and Russia. Introduced in N America.

Biology. The primary hosts are *Lonicera xylosteum* and *L. tatarica.* The leaves of the infested shoots are curled and get yellow spots in spring. As grown nymphs the fundatrigeniae move from the leaf rolls to the twigs. They are easily observed due to the rich wax production. When they become alate they fly to *Picea abies,* where the virginogeniae feed on the roots.

Genus *Mimeuria* Börner, 1952

Mimeuria Börner, 1952: 197, 259.
Type-species: *Neorhizobius ulmiphilus* del Guercio, 1917.
Survey: 284.

In several respects similar to *Prociphilus,* but virginogenia with one -segmented tarsi, and its first instar with short, erect empodial hairs (Fig. 184).
One species in the genus.

Mimeuria ulmiphila (del Guercio, 1917)
Figs. 181–187.

Neorhizobius ulmiphilus del Guercio, 1917: 251. – Survey: 284.

Fundatrix. Olive greyish green, covered with white wax wool. Body with rather few short dorsal hairs, longer and more numerous towards the end of abdomen. Head with 3 pairs of wax gland plates, sometimes with 3–4 smaller plates replacing the middle pair; prothorax with 4 spinal (sometimes fused) and 2 marginal plates; meso- and metathorax and abd. segm. I–VI each with 6 plates; VII with 4 plates, VIII with 2 plates (Fig. 181). Antenna 5-segmented, 0.14–0.18 × body; segm. III longer than V, shorter than IV+V; primary rhinaria subcircular, surrounded by short hairs. Apical segm. of rostrum 0.6–0.7 × 2sht. Siphuncular pores absent. 3.5–4.5 mm.

Alate viviparous female from primary host (fundatrigenia). Abdomen olive green. Head without distinct wax gland plates; prothorax with 1–2 small plates; mesothorax without plates; metathorax with 0–2 spinal plates; abd. segm. I–III with 2 spinal and 2 marginal plates, IV–VI also with more or less distinct pleural plates, VII with 4 plates, VIII with 2. Antenna 6-segmented, 0.3 × body; segm. III about 4 × II, abous as long as IV+V; IV, V, and VI about equally long; processus terminalis shorter than 0.2 × VIa; primary rhinaria of irregular shape, surrounded by spinules; secondary rhinaria narrow, half-ring-shaped, with dotted borders (Fig. 183); segm. III with 10–15, IV with 4–7, V with 3–5, VI without secondary rhinaria. Apical segm. of rostrum 0.5–0.7 × 2sht., with 4 accessory hairs and rows of spinules. First tarsal segm. with 2 hairs. Media of fore wing unbranched. Siphuncular pores present. 2.6–3.3 mm.

Apterous viviparous female on secondary host (virginogenia). Amber yellow, covered with white wax wool. Body shape globular. With 6 longitudinal rows of reticulate wax gland plates, more distinct in immature specimens than in adults, abd.

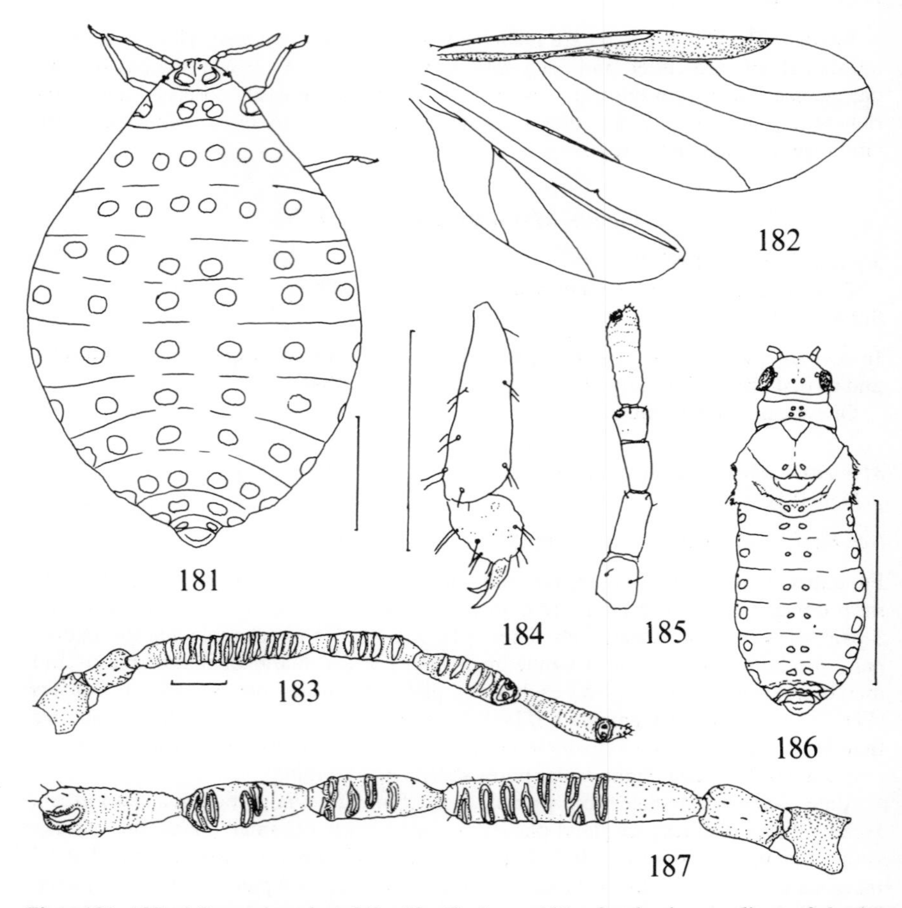

Figs. 181–187. *Mimeuria ulmiphila* (d. Gu.). – 181: fundatrix, outline of body (dorsal view) and arrangement of wax gland plates. (Scale 1 mm); 182: alate fundatrigenia, wings. (Scale as 181); 183: alate fundatrigenia, antenna. (Scale 0.1 mm); 184: virginogenia, tibia and tarsus of middle leg of first instar nymph. (Scale 0.1 mm); 185: virginogenia, antenna of adult specimen; 186: sexupara, arrangement of wax gland plates on body segments. (Scale 1 mm); 187: antenna of sexupara. (181–184 after Krzywiec, 185–187 after Marchal; redrawn).

segm. VII with 4 plates, VIII with 2; head with several plates; wax gland plates frequently divided into small subgroups of facets bounded by distinct internal borders. Antenna 5-segmented, very short; segm. V longer than II, which is about as long as III+IV; processus terminalis vestigial (Fig. 185); primary rhinaria surrounded by short hairs. Rostrum reaching past hind coxae. Tarsi one-segmented through absorption of segm. I into the base of segm. II. Less than 2 mm.

Alate viviparous female from secondary host (sexupara). Wax gland plates well developed, especially on posterior part of abdomen (Fig. 186), often divided into subgroups of facets by well marked internal borders, the subgroups now and then being more or less concentric; head with 2 plates, often smaller than ocelli; prothorax with 4 small spinal plates, mesothorax with 2, metathorax with 2, abd. segm. I–VI each with 2 spinal and 2 marginal plates, VI also with 2 very small pleural glandular areas; plates on VII and VIII fused into transversal glandular areas. Antenna rather similar to that of fundatrigenia; processus terminalis vestigial, not projecting more than 0.1 mm beyond the primary rhinarium (Fig. 187); primary and secondary rhinaria narrow, half-ring-shaped, with short-haired or dotted borders; segm. III with about 7, IV with about 5, V with about 3, VI without secondary rhinaria. Venation of wings as in fundatrigenia. Siphuncular pores absent. About 1.9 mm.

Distribution. Not found in Scandinavia, but may be overlooked. – Widespread in Europe (Britain, Germany, Poland, USSR, France, Italy).

Biology. The species causes leaf-nests on the primary hosts, *Acer campestre* and *A. platanoides,* and are here visited by ants (Krzywiec 1962). The feeding of the aphids causes the petioles to be twisted so that the leaves are turned at an angle of 180°, contiguous to petiole, and the shoots are twisted and shortened (Shaposhnikov 1964). Alate fundatrigeniae can be found on *Acer* from mid-summer to October or November in Poland. The virginogeniae live on roots of *Ulmus* (and perhaps also other woody plants) enveloped by brown mycorrhizal cysts all year round. Sexuparae are produced in autumn. The generations on *Acer* have only been observed a few times, so the species may be predominantly anholocyclic. It is probably very rare. The colonies on the secondary hosts are easily overlooked, but, nevertheless, are more frequently found than the generations on the primary hosts.

Note. The description is based on Krzywiec (1962), who discovered that *M. ulmiphila* uses *Acer* as the primary host and is identical with *Paraprociphilus ucrainensis* Mamontova, and Marchal (1933), who described virginogeniae and sexuparae in detail. The genus was placed in Pemphigini by Börner because the empodial hairs of newborn virginogeniae are short and erect. After the discovery of the generations on *Acer,* Krzywiec placed it in Pachypappini (= Prociphilini) to which *Paraprociphilus* Mordvilko belongs, and even suggested that *Mimeuria* might be ascribed rank of subgenus in the genus *Paraprociphilus* when detailed morphological analysis of non-European *P.* spp. could be carried out. *Acer* is also the primary host of the American *P. tesselatus* (Fitch).

Eastop & Hille Ris Lambers (1976) regard *Paraprociphilus* as a subgenus of *Prociphilus.*

TRIBE PEMPHIGINI

Newborn virginogeniae with very short, uncurved empodial hairs, which are shorter than the claws. Galls are produced in spring on leaves or petioles of *Populus*. The secondary hosts of most species are herbaceous plants. Media of fore wing unbranched. Apical segm. of rostrum with pale subapical zone.

Genus *Thecabius* Koch, 1857

Thecabius Koch, 1857: 294.
Type-species: *Thecabius populneus* Koch, 1857.
= *Pemphigus affinis* Kaltenbach, 1843.
Survey: 421.

Fundatrix with 5- or 6-segmented antennae; other viviparous female morphs usually with 6-segmented antennae (seldom 5-segmented in apterae). All morphs have wax gland plates, in apterous viviparous females placed in 4–6 longitudinal rows; fundatrix and sometimes also apterous viviparous females on secondary hosts not only have plates on body, but also on head (Fig. 188). Alate fundatrigeniae with 2–5 hairs on first tarsal segments, with 3 or more hairs on at least some tarsi.

There are 14 species in the world. The genus is subdivided into two subgenera.

Key to subgenera of *Thecabius*

Fundatrices

1 Apical segm. of rostrum blunt, with 2 accessory hairs. *Thecabius* Koch s. str. (p. 167)

– Apical segm. of rostrum pointed, without accessory hairs. *Parathecabius* Börner (p. 169)

Alate viviparous females from primary hosts (fundatrigeniae)

1 Secondary rhinaria narrower than spaces between adjacent rhinaria. Mesonotum rather often with two wax gland plates. .... *Thecabius* Koch s. str. (p. 167)

– Secondary rhinaria broader than spaces between adjacent rhinaria. Mesonotum without wax gland plates. ..... *Parathecabius* Börner (p. 169)

Apterous viviparous females on secondary hosts (virginogeniae)

1 Wax gland plates present on head. Apical segm. of rostrum with 4–6 accessory hairs. .... *Thecabius* Koch s. str. (p. 167)

– Wax gland plates absent from head. Apical segm. of rostrum without accessory hairs. .... *Parathecabius* Börner (p. 169)

Alate viviparous females from secondary hosts (sexuparae)

1 Ant. segm. III usually with more than 12 secondary rhinaria. Head with wax gland plates. .... *Thecabius* Koch s. str. (p. 167)

– Ant. segm. III with 6–12 secondary rhinaria. Head without wax gland plates. .... *Parathecabius* Börner (p. 169)

Subgenus *Thecabius* Koch, 1857 s. str.

Apical segment of rostrum rather blunt, in fundatrix shorter than 0.7 × 2sht. and with the apical angle just over 90°; in apterae on secondary hosts with 4–6 accessory hairs, in fundatrix with 2 such hairs. Head with wax gland plates except in alate fundatrigeniae; the alate fundatrigenia usually with plates on mesonotum. Primary rhinaria surrounded by short hairs.

Seven species in the world, one species in Scandinavia.

35. ***Thecabius (Thecabius) affinis*** (Kaltenbach, 1843)
Figs. 188–191.

Pemphigus affinis Kaltenbach, 1843: 182. – Survey: 421.

Fundatrix. Green or bluish green, covered with wax. Head and appendages black. Body oval, with numerous very short hairs. Head with several well developed wax gland plates; body with 6 longitudinal rows of plates. Antenna 5- or 6-segmented, 0.20–0.25 × body; segm. III about twice as long as IV; II about 1.9–2.0 times as long as its greatest width. Apical segm. of rostrum about 0.55–0.6 × 2sht., with 2 accessory hairs. Siphuncular pores absent. 3.4–4.5 mm.

Alate viviparous female from primary host (fundatrigenia). Abdomen greenish. Head usually without wax gland plates; pronotum and abd. segm. I–VII with small spinal and marginal plates, I–V sometimes also with pleural plates; VIII with 2 plates; both meso- and metanotum usually have one pair of spinal plates. Antenna about 0.5 × body; segm. III with thickened basal part and considerably longer than IV+V, about twice as long as VI; V longer than IV and as long as or shorter than VI; processus terminalis about 0.25 × VIa; secondary rhinaria narrower than spaces between adjacent rhinaria; on III: 15–30, IV: 4–9, V: 4–11, VI: 0–7. First tarsal segm. usually with 5 hairs. Siphuncular pores present, small and inconspicuous. 2.2–3.1 mm.

Apterous viviparous female on secondary host (virginogenia). Dirty yellowish white or yellowish green, with shining wax threads. Head with several wax gland plates; abd. segm. III–VI each with 2 spinal, 2 pleural, and 2 marginal transverse oval plates; VII with 2 plates. Antenna 6-segmented, about 0.3 × body; segm. III about as long as IV+V, with 0–2 secondary rhinaria. Rostrum reaching past hind coxae; apical segm. 0.11–0.13 mm long, about 0.5 × 2sht., with 4–6 accessory hairs. Siphuncular pores absent. 2.1–2.9 mm.

Alate viviparous female from secondary host (virginopara and sexupara). Similar to fundatrigenia, but head with wax gland plates, and siphuncular pores absent. Dorsum with several rows of wax gland plates, some of which have very small facets. Secondary rhinaria: III: 9–22, IV: 2–7, V: 0–6 (in sexuparae at most 2), VI: 0. Ant. segm. V about as long as VI in virginoparae (containing embryos with rostrum), distinctly shorter than VI in sexuparae (containing embryos without rostrum) (Fig. 190). 1.9–2.9 mm.

Distribution. Very common and widespread in Denmark; in Sweden from Sk. north to P. Lpm., common; in Norway common and widespread, north to NTy; in Finland known from Ab, N, Ta, and Oa. – Europe and Asia, from the British Isles east to Siberia and Mongolia, south to Portugal, Spain, Bulgaria, Turkey, and Caucasus; common in N Germany, N Poland, and NW Russia. N Africa. Introduced into N America and now widespread.

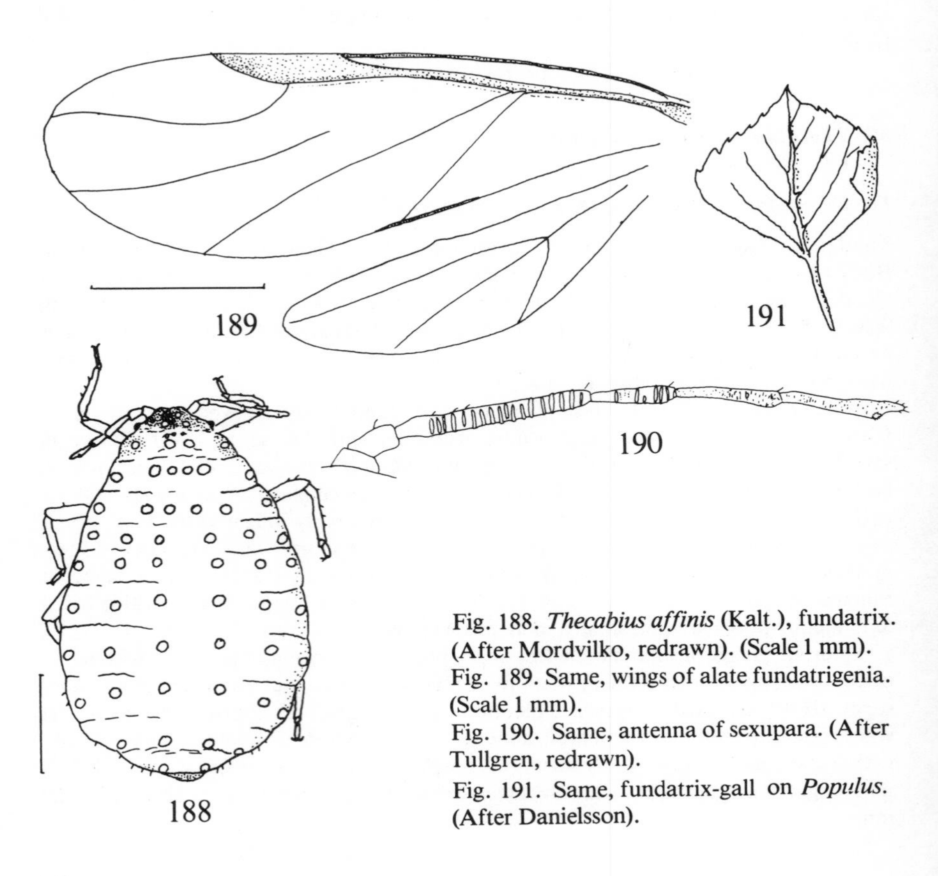

Fig. 188. *Thecabius affinis* (Kalt.), fundatrix. (After Mordvilko, redrawn). (Scale 1 mm).
Fig. 189. Same, wings of alate fundatrigenia. (Scale 1 mm).
Fig. 190. Same, antenna of sexupara. (After Tullgren, redrawn).
Fig. 191. Same, fundatrix-gall on *Populus*. (After Danielsson).

Biology. The species is host-alternating between *Populus (nigra, italica)* (I) and *Ranunculus (repens, acer, bulbosus, flammula* a. o.) (II). Fundatrix lives in a small gall formed by sharp folding of a part of the edge of the leaf (Fig. 191), especially of young poplars. The fundatrigeniae leave this fundatrix-gall through a narrow opening between the folded part and the underside of the leaf and move to the mid rib, especially of young leaves. Their feeding causes folding of the leaves along the mid ribs towards the underside of the leaves so that another kind of gall arises. The roof-like upper side of the folded leaf becomes rough and blistered and gradually turns reddish. Alate fundatrigeniae fly to *Ranunculus*, according to Mordvilko (1935) in the second half of June and the beginning of July in the Leningrad area, according to Tullgren (1909) and Danielsson (1976) in the middle of July in central and southern Sweden. Several generations of virginogeniae live here on stem bases, runners, and undersides of leaves. Alate virginoparae are produced in the colonies on *Ranunculus* in summer and autumn and fly to other *Ranunculus* plants. Alate sexuparae occur from the end of August and fly from *Ranunculus* to *Populus* to bear the sexuales (which have 4-segmented antennae) in autumn. Overwintering of immature virginogeniae on *Ranunculus* has been observed by Mordvilko (1935). The overwintered colonies can produce alate virginoparae early in the year and alate sexuparae from August.

Subgenus *Parathecabius* Börner, 1950

Parathecabius Börner, 1950: 18.
Type-species: *Thecabius lysimachiae* Börner, 1916.
Survey: 336.

Apical segm. of rostrum rather pointed, in fundatrix longer than 0.7 × 2sht. and with the apical angle about 50–60°; apterae without accessory hairs; such hairs usually also absent from alatae. Head without wax gland plates (except in fundatrix); alate fundatrigenia without plates on mesonotum. Primary rhinarium on ant. segm. V of fundatrigenia not surrounded by hairs.

With 6 or 7 species in the world, one species in Scandinavia. The secondary host is *Lysimachia.* Another little-known species, which may be found here in future on *Primula* spp. in gardens or indoors, is shortly treated below. It seems impossible to separate these two species morphologically, so a key cannot be given.

36. ***Thecabius (Parathecabius) lysimachiae*** Börner, 1916
Figs. 192, 193.

Thecabius lysimachiae Börner in Börner & Blunck, 1916: 31, 38.
Survey: 422.

Fundatrix. Most differences between fundatrices of this species and *affinis* are given in the description of the subgenus. Ant. segm. II about 1.6 as long as its greatest width. Apical segm. of rostrum about 0.85 × 2sht.

Alate viviparous female from primary host (fundatrigenia). Abdomen dirty greenish, almost black (as a nymph yellowish green). Head and mesonotum without wax gland plates; abdomen with more or less well developed spinal, pleural, and marginal plates, often absent from several segments. Antenna about 0.4–0.5 × body; segm. III a little longer than IV+V; processus terminalis 0.2–0.25 × VIa; secondary rhinaria narrow, a little wider than spaces between adjacent rhinaria, on III: 15–21, IV: 4–10; V: 5–9, VI: 8–12. Apical segm. of rostrum about 0.65 × 2sht., rather pointed, without accessory hairs. First tarsal segm. usually with 3-2-2 hairs. Siphuncular pores inconspicuous. 2.3–2.6 mm.

Apterous viviparous female on secondary host (virginogenia). Brownish or greyish green, with wax. Head without wax gland plates; abd. segm. III–VI each with 2 spinal and 2 pleural plates; VII with 2 plates. Antenna usually 6-segmented, rarely 5-segmented; segm. VI longer than other segments; III distinctly shorter than IV+V, without rhinaria. Apical segm. of rostrum short and broad, about 0.08–0.13 mm long, 0.6–0.8 × 2sht. 1.3–1.8 mm.

Alate viviparous female from secondary host (sexupara). With fewer secondary rhinaria than fundatrigenia, narrow, stripeshaped, on III: 6–12, IV: 2–4, V and VI: 0. Ant. segm. III as long as or shorter than IV+V. Apical segm. of rostrum about 0.65 × 2sht., with 0–1 accessory hairs. First tarsal segm. with 4 or 5 hairs in large, 2 or 3 hairs in small, specimens. 2.1–2.4 mm.

Distribution. In Denmark known from the Copenhagen area (NEZ) and Horsens (EJ) on potted *Lysimachia nummularia;* in Sweden known from Sk. and Upl.; not in Norway and Finland. – Europe: British Isles, France, Germany, Austria, and Russia; not in N Germany or Poland.

Biology. The primary host is *Populus nigra,* the secondary *Lysimachia nummularia.* The fundatrix gall is oval, rather thick-walled and placed on the upper side of the leaf, but not along the mid rib. The fundatrigeniae produce another kind of gall by folding the leaf in an irregular way (Fig. 193). The generations on *Populus* are not commonly found (never observed in Denmark), but the reason may be that they easily escape

Fig. 192. *Thecabius (Parathecabius) lysimachiae* Börner, antenna of alate fundatrigenia. (Drawn after photo in Furk & Prior).

Fig. 193. Same, gall produced by fundatrigeniae on *Populus.* (After Danielsson).

notice because they occur in the upper parts of the trees. Anholocyclic overwintering on runners and roots of the secondary host seems to be common and may be the normal way of overwintering in W & N Europe. The species can be a pest to *Lysimachia nummularia* in gardens.

Thecabius (Parathecabius) auriculae (Murray, 1877)

Trama auriculae Murray, 1877: 570. – Survey: 422.

Apterous viviparous female (virginogenia). Covered with white wax powder and wool, mostly on posterior part of body. Abdominal wax gland plates as in virginogenia of *lysimachiae*. Antenna 5-segmented. Rostrum reaching past hind coxae. 1.3–1.5 mm.

Alate viviparous female. Abdomen dark green. Distinct wax gland plates present on all thoracic and abdominal segments. Antenna similar to that of sexupara of *lysimachiae*. About 2.5 mm.

Distribution. Not in Scandinavia, but perhaps overlooked. – Great Britain; Japan.

Biology. Colonies of apterous virginogeniae occur on roots and root collars of *Primula* (*P. auricula, P.* sp.). The species is regularly a pest to auriculas in England. It especially attacks potted plants, the leaves of which become yellowish and deformed. Alatae are extremely rare.

Note. The information above is taken from Theobald (1929), Börner & Heinze (1957), and Stroyan (1964a).

Genus *Pemphigus* Hartig, 1839

Pemphigus Hartig, 1839: 645.
Type-species: *Aphis bursaria* Linné, 1758.
Survey: 340.

Fundatrix with 4-segmented antennae (but segm. III sometimes with incipient division); other viviparous morphs with 5- or 6-segmented antennae. Fundatrix with spinal, pleural, and marginal wax gland plates on most body segments; apterous virginogeniae with spinal and pleural plates; alatae sometimes with spinal plates only on a few segments, but occasionally also with marginal plates on most segments and then with only a few spinal or pleural plates; alate fundatrigeniae without plates on mesonotum; wax gland plates absent from head (with some few exceptions). Transverse oval, often rather narrow and almost ringlike secondary rhinaria present on ant. segm. III–VI in fundatrigeniae of most species, on III and IV in sexuparae (on III–IV(–V) in fundatrigeniae of *immunis* and *populi*). Primary rhinarium on ultimate ant. segm. surrounded by short hairs; primary rhinarium on penultimate ant. segm. of fundatrigeniae usually 2–3 times as broad as secondary rhinaria and not surrounded by hairs; primary rhinarium on penultimate ant. segm. of sexuparae and apterae surrounded by hairs and not enlarged. Apical segment of rostrum usually without accessory hairs. First tarsal segm. with 2(3)-2-2 hairs. Siphuncular pores small or absent , always absent from fundatrices, apterous virginogeniae, and sexuparae.

With 58 species in the world, 9 species in Scandinavia. Eleven species are described below, mainly based on literary information, especially from Stroyan (1964) and Furk & Prior (1975). I have not examined all species, nor all morphs described. The intraspecific variation is large, and the given range may be too narrow. Therefore, caution is advised when using the keys, also because additional species may occur. It is a taxonomically difficult genus. The species are more safely separated on biological than on morphological characters.

Most species are dioecious and produce solid galls on leaves, petioles, or branches of the primary hosts, *Populus* spp. of the *nigra*-group *(Eupopulus)*, in spring and early summer. They are not visited by ants.

Key to species of *Pemphigus*

Fundatrices and galls
(except *fuscicornis* and *saliciradicis)*

1 Galls placed directly on branches (Figs. 195, 217). Fundatrix usually larger than 2.8 mm. 2

– Galls placed on leaves or petioles. Fundatrix usually smaller than 2.8 mm. 3

2 (1) Ant. segm. III about 1.1–1.2 × IV. On *Populus nigra* or *P. italica*. .... *trehernei* Foster

– Ant. segm. III 1.3–1.6 × IV. Not on *Populus nigra* or *P. italica*. .... 37. *borealis* Tullgren

3 (1) Galls on petioles. 4

– Galls on uppersides of leaves on the midrib. 6

4 (3) Galls not spirally twisted (Fig. 200 A). .... 38. *bursarius* (Linné)

– Galls spirally twisted (Figs. 200 B, 206). 5

5 (4) Ant. segm. III shorter than IV. Galls with apterous fundatrigeniae and later alate sexuparae, not emptied until autumn. 45. *spyrothecae* Passerini

– Ant. segm. III longer than IV. Galls with alate fundatrigeniae, emptied during summer. .... 43. *protospirae* Lichtenstein

6 (3) Gall blister-shaped, oval or nearly globular, often irregularly shaped, rather narrow at base, placed on the midrib close to the base of the leaf (Fig. 205). Ant. segm. III about 0.95 × IV. .... 41. *populi* Courchet

– Gall formed as a blister-shaped, thick-walled swelling of the midrib, usually near the middle of the leaf (Figs. 200 C, 204). Ant. segm. III usually longer than 0.95 × IV. 7

7 (6) Pronotum with at most one, but often without spinal wax gland plates; usually without well-defined pigmented sclerotized area, or with only irregular, small, scattered spots. Gall spindle-shaped and a little wrinkled, similar to that of *phenax*. .... *gairi* Stroyan

– Pronotum often with well developed spinal wax gland plates and a more or less distinctly sclerotized area. Gall spindle-shaped and a little wrinkled, or rather broad, oval, and smooth. 8

8 (7) Fundatrix green, also as a young nymph. Gall rather broad and smooth, of a dull reddish colour without much yellow tinge (Fig. 200 C). .... 42. *populinigrae* (Schrank)

– Fundatrix green as an adult, red as a young nymph. Gall elongate, spindle-shaped, a little wrinkled, and often tinged with yellow laterally (Fig. 204). .... 40. *phenax* Börner & Blunck

Apterous viviparous females (not fundatrices)

1 Siphuncular pores present. In spirally twisted galls on leaf petioles of *Populus*. .... 45. *spyrothecae* Passerini

– Siphuncular pores absent. Not in galls on *Populus*. 2

2 (1) Wax gland plates present only on abd. segm. (IV–)V–VII. On Leguminosae. .... 41. *populi* Courchet

– Wax gland plates usually present on abd. segm. III–VII. Not on Leguminosae. 3

3 (2) On *Bidens* (nymphs; the adults are alate). .... 37. *borealis* Tullgren

– Not on *Bidens*. 4

4 (3) Apical segm. of rostrum 1.2–1.4 × 2sht. On *Filago* and *Gnaphalium*. .... 42. *populinigrae* (Schrank)

– Apical segm. of rostrum shorter than 1.2 × 2sht. Not on *Filago* and *Gnaphalium*. 5

5 (4) Head often with two wax gland plates. Spinules present between the long hairs on distal margin of first segm. of hind tarsus and on the sole of 2sht. On roots of *Salix*. .... 44. *saliciradicis* (Börner)

– Head without wax gland plates. Distal margin of first segm. of hind tarsus and the sole of 2sht. sometimes with imbrications, not with distinct spinules. Not on *Salix*. 6

6 (5) On Umbelliferae. 7

– Not on Umbelliferae. 9

7 (6) On carrots (*Daucus carota*). .... 40. *phenax* Börner & Blunck

– Not on carrots. 8

8 (7) Reddish or yellowish. Apical segm. of rostrum 0.08–0.095 mm. On *Aethusa*. .... *gairi* Stroyan

– Yellowish or greenish. Apical segm. of rostrum 0.10–0.13 mm. On *Berula, Sium,* and *Apium.* .... 43. *protospirae* Lichtenstein
9 (6) Greyish green. On *Matricaria* (incl. *Tripleurospermum*) (and perhaps *Beta* and *Chenopodium*). .... 39. *fuscicornis* (Koch)
– Yellowish or greenish white. On other plants. 10
10 (9) On *Aster tripolium.* .... *trehernei* Foster
– Not on *Aster tripolium.* .... 38. *bursarius* (Linné)

Alate viviparous females
(except *fuscicornis*)

1 Primary rhinarium on penultimate ant. segm. not surrounded by short hairs (fundatrigeniae). 2
– Primary rhinarium on penultimate ant. segm. surrounded by short hairs (sexuparae). 10
2 (1) Ultimate ant. segm. without secondary rhinaria, penultimate segm. with 0–2 (Fig. 208). Siphuncular pores absent. 41. *populi* Courchet
– Ultimate ant. segm. with (0–)1–11 secondary rhinaria; if without secondary rhinaria, then penultimate segm. with more than one secondary rhinarium. Siphuncular pores present. 3
3 (2) Primary rhinarium on penultimate ant. segm. very broad, rarely narrower than twice the basal diameter of the segment, often covering almost its half length (Fig. 201). 40. *phenax* Börner & Blunck
– Primary rhinarium on penultimate ant. segm. not very broad, rarely broader than twice the basal diameter of the segment, never covering its half length. 4
4 (3) Apical segm. of rostrum 0.10–0.12 mm, 0.5 × 2sht. or longer. Ant. segm. III longer than IV+V. 5
– Apical segm. of rostrum 0.07–0.10 mm, 0.5 × 2sht. or shorter. Ant. segm. III 0.7–1.2 × IV+V. 7
5 (4) Basal 1/5–1/4 of ant. segm. III (the part proximal of the small tooth-like process on basal part) devoid of rhinaria. Processus terminalis well defined, finger-shaped. Primary rhinarium on penultimate ant. segm. much broader than the secondary rhinaria (Fig. 202). .... 43. *protospirae* Lichtenstein
– Ant. segm. III with rhinaria also on basal 1/5. Processus terminalis not well defined, rather thick at least at base, or primary rhinarium on penultimate ant. segm. not much broader than the secondary rhinaria. 6
6 (5) Secondary rhinaria narrower than primary rhinarium on penultimate ant. segm. (Fig. 209). Ant. segm. V of about the same length as IV. .... 42. *populinigrae* (Schrank)

– Secondary rhinaria almost as broad as primary rhinarium on penultimate ant. segm. (Fig. 194). Ant. segm. V 1.1–1.2 × IV. .... 37. *borealis* Tullgren

7 (4) Ant. segm. III less than twice as long as IV (Fig. 203). .... *gairi* Stroyan

– Ant. segm. III more than twice as long as IV. 8

8 (7) Secondary rhinaria on segm. V+VI of both antennae number to 5–23 (Fig. 196). .... 38. *bursarius* (Linné)

– Secondary rhinaria on segm. V+VI of both antennae number to 24–36. 9

9 (8) Ant. segm. III 2.5–3.0 × IV. Primary rhinarium on segm. V relative large (Fig. 194). .... 37. *borealis* Tullgren

– Ant. segm. III 2.2–2.6 × IV. Primary rhinarium on segm. V relatively small (Fig. 215). .... *trehernei* Foster

10 (1) Spinal wax gland plates present on abd. segm. I–V(–VI) and VIII, paired or single. 11

– Spinal wax gland plates present only on abd. segm. I–III (–IV), usually paired. (The identification of the following species is most uncertain). 13

11 (10) Apical segm. of rostrum pointed, longer than 2sht. ... 42. *populinigrae* (Schrank)

– Apical segm. of rostrum blunt, shorter than 2sht. 12

12 (11) Ant. segm. III shorter than IV+V. .... 44. *saliciradicis* (Börner)

– Ant. segm. III longer than IV+V. .... 37. *borealis* Tullgren

13 (10) Ant. segm. VI as long IV+V or longer, longer than 1.6 × V. 14

– Ant. segm. VI frequently shorter than IV+V, usually shorter than 1.8 × V. 15

14 (13) Apical segm. of rostrum 0.12–0.14 mm. Ant. segm. VI about 1.7–2.1 × V. .... 41. *populi* Courchet

– Apical segm. of rostrum 0.07–0.09 mm. Ant. segm. VI about 2.0–2.8 × V. .... *trehernei* Foster

15 (13) Genital plate with two dark patches. .... 45. *spyrothecae* Passerini

– Genital plate with one narrow transverse dark patch. 16

16 (15) Genital plate with more than 20 hairs, usually with 12–21 hairs around the posterior margin. Apical segm. of rostrum with 0–2 accessory hairs. .... 43. *protospirae* Lichtenstein

– Genital plate with less than 20 hairs, usually with 5–10 hairs around the posterior margin. Apical segm. of rostrum without accessory hairs.
38. *bursarius* (Linné), *gairi* Stroyan, or 40. *phenax* Börner & Blunck

37. ***Pemphigus borealis*** Tullgren, 1909
Figs. 194, 195.

Pemphigus borealis Tullgren, 1909: 142. – Survey: 340.

Fundatrix. Very similar to *bursarius*. Ant. 0.16–0.19 × body; segm. III about 1.3–1.6 × IV. 1.6–3.2 mm.

Alate viviparous female from primary host (fundatrigenia). Abdomen bluish, yellowish or greyish green. Wax gland plates absent from some abd. segm. Antenna 6-segmented, about 0.5 × body; segm. III rather long, 2.5–3.0 × IV, longer than IV+V; V slightly longer than IV and about 0.6–0.8 × VI; primary rhinarium on V rather similar to the secondary rhinaria, which are relatively large, ringlike, and uniform (Fig. 194); secondary rhinaria on III: 13–19, IV: 4–5, V: 6–9, VIa: 6–10. Apical segm. of rostrum (0.08–)0.10–0.11 mm, about 0.5 × 2sht. Most other characters as in *bursarius*. 1.5–2.6 mm.

Alate viviparous female from secondary host (sexupara). Rather similar to *trehernei*, but with more secondary rhinaria, on segm. III: 8–10, IV: 2–3; ant. segm. III longer than IV+V, and VI shorter than IV+V. About 2 mm.

Distribution. Eastern part of Sweden: Ög., Sdm., Upl., Gstr.; southern part of Finland: Ab, N, Ka, St, Ta; not in Denmark or Norway. – Britain, Germany, Poland, Russia, Siberia, Mongolia. A record from Iran (Davatchi 1948) probably refers to *P. immunis*, according to Bodenheimer & Swirski (1957). Gittins & al. (1976) recorded it from N America (Idaho). It has been suggested (Shaposhnikov 1964) that the origin is the Far East, but I have not seen any published records from that region.

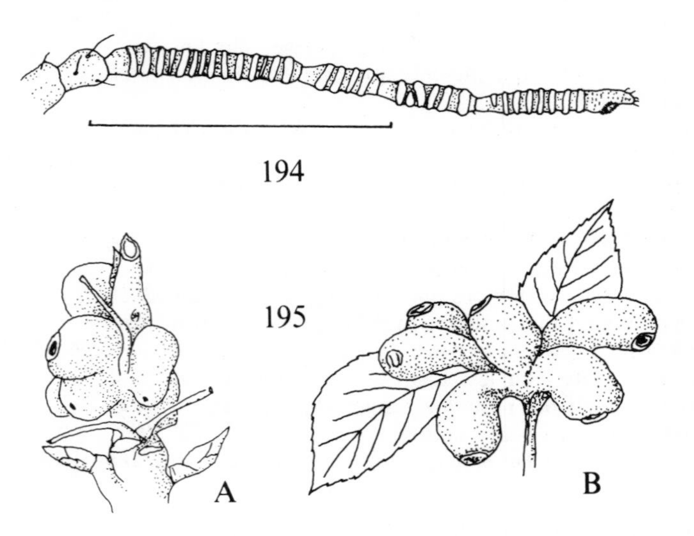

Figs. 194, 195. *Pemphigus borealis* Tullgr. – 194: antenna of alate fundatrigenia. (Scale 0.5 mm); 195 A & B: galls on branch of *Populus*. (194 after Tullgren, 195 A after Danielsson, 195 B after Theobald; redrawn).

Biology. The galls are oval, green, of hazel-nut size, and usually placed in groups of five or six at the base of branches of *Populus*, in Sweden on *P. laurifolia* and *P. balsamifera*. They are growths from the proper branch, not from the leaves or petioles (Fig. 195). Each gall has apically a more or less narrow opening, which usually is turned downwards. Dolgova (1970) found that all fundatrigeniae are alate and leave the galls between mid June and mid August in Russia. Tullgren found fundatrix together with alate fundatrigeniae in galls at the end of August in Sweden. The fundatrigeniae fly to *Bidens (tripartita, cernua)*, where only one generation is developed, viz. the alate sexuparae. These return to *Populus* in autumn (Dolgova 1970).

38. ***Pemphigus bursarius*** (Linné, 1758)
Plate 1: 9. Figs. 196–199, 200 A.

Aphis bursaria Linné, 1758: 453. – Survey: 341.

Fundatrix. Greyish green, slightly wax powdered. Head and legs brown. Wax gland plates as described above of the genus; pronotum with spinal and marginal plates. Antenna 4-segmented, 0.12–0.15 × body; segm. III as long as or longer than IV. Apical segm. of rostrum about 0.10 mm, 0.7–0.9 × 2sht. Siphuncular pores absent. 2.4–3.0 mm.

Alate viviparous female from primary host (fundatrigenia). Abdomen greyish green or brownish green, slightly powdered. Wax gland plates absent from thorax, may be present on abdomen, but less developed than in nymph. Antenna (Fig. 196) of 6 (seldom 5) segments, 0.33–0.40 × body; segm. III more than twice as long as IV and 0.8–1.2 × IV+V; processus terminalis well-defined, about 0.17–0.20 × VIa; primary rhinaria on V (in 6-segmented antenna) not very large, rarely broader than 2 × basal diameter of segment, not surrounded by short hairs; secondary rhinaria on III: 8–17, IV: 2–9, V: 2–8, VI: 2–10 (VI rarely without secondary rhinaria in specimens from autumn); III with small tooth-like process near base, usually distal of the proximal secondary rhinarium. Apical segm. of rostrum 0.08–0.10 mm, 0.4–0.5 × 2sht., often with slightly convex margins. Wing veins shadowed with brown. Siphuncular pores present, but small. 1.6–2.5 mm.

Apterous viviparous female on secondary host (virginogenia). Yellowish white, with white tuft of wax on the posterior part of abdomen. Wax gland plates present on abd. segm. III–VII, usually 4 plates on each of the segm. III–VI (but the anterior plates may be absent), 2 plates on VII (Fig. 197). Antenna 5- or 6-segmented, about 0.2 × body; segm. VI longer than IV+V in 6-segmented antenna; processus terminalis about 0.25–0.33 × VIa (Va). Apical segm. of rostrum 0.09–0.12 mm (longer in hibernating virginoparae than in summer specimens), about 0.8 × 2sht. Siphuncular pores absent. 1.5–2.7 mm.

Alate viviparous female from secondary host (sexupara). Abdomen brownish orange or (in old individuals) greyish green. Wax gland plates present both on thorax and abdomen, but the abdominal plates may be difficult to observe. Antenna 6-segmented, about 0.33 × body; segm. III of about the same length as IV+V or shorter (Fig. 199); VI frequently shorter than IV+V; processus terminalis about 0.25 × VIa; primary rhinaria surrounded by rings of short hairs; III with 5–10 secondary rhinaria on distal 1/2–3/4, V with (1–)2–4 on distal half, V and VI without secondary rhinaria. Apical segm. of rostrum about 0.11 mm, 0.6–0.7 × 2sht., without accessory hairs. Siphuncular pores absent. 1.6–2.7 mm.

Distribution. In Denmark very common and widespread; common in Sweden north to Upl.; widespread in the southern part of Norway, north to Bv and HEs; in Finland north to Sb. – Common and widespread all over Europe, south to Portugal, Spain, Italy, and Yugoslavia, including the British Isles, N Germany, N Poland, and NW & W Russia. N Africa: Morocco, Algeria, Tunisia, Egypt. Asia: Turkey, Lebanon, Syria, Iraq, Iran, Siberia, C Asia, Mongolia, E Himalayas; records from China are doubtful. Also Australia, N Zealand, and N & S America.

Biology. The species is host-alternating between *Populus (nigra, italica)* (I) and roots of various herbs (II) especially Compositae, e.g. *Crepis, Lactuca, Lapsana, Sonchus, Taraxacum,* and *Tussilago.* Purse- or pear-shaped, yellowish or reddish galls are formed on the petioles of the primary host (Fig. 200 A). There may be more than one gall on the same petiole. In summer, often rather late, the alate fundatrigeniae leave the gall through a lateral opening. In Denmark I have observed galls full of alatae (fun-

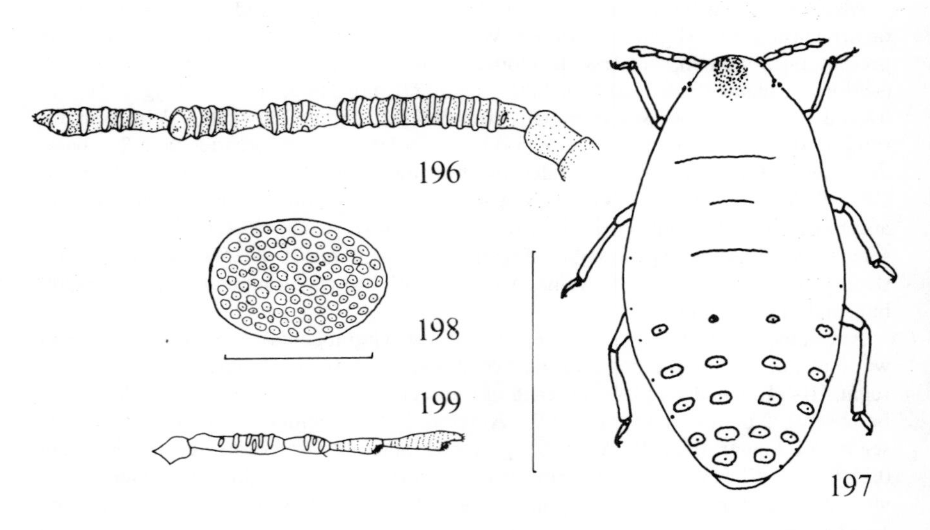

Figs. 196–198. *Pemphigus bursarius* (L.). – 196: antenna of alate fundatrigenia; 197: apterous virginogenia from root of *Tussilago.* (Scale 1 mm); 198: wax gland plate of apterous virginogenia. (Scale 0.1 mm). (196 drawn after photo in Furk & Prior, 198 after Zwölfer; redrawn).

Fig. 199. Same, antennal segments II–VI of sexupara. (After Tullgren, redrawn).

datrigeniae) in early September (Gråsten, SJ), and Tullgren (1909) made the same observation in late September (Stockholm, Upl.). Colonies of apterous virginogeniae founded on roots of composites in summer produce sexuparae. They fly to *Populus* in September–October to bear the small sexuales. Fundatrigeniae produced in the galls in autumn may be lost. The biology is still rather imperfectly known; more than one species may be involved. Virginogeniae found on roots of various plants are very similar. The parthenogenetic reproduction may continue for years. Anholocyclic overwintering of apterous virginogeniae on the secondary hosts has been observed after the sexuparae had left the colonies (Tullgren 1909, Mordvilko 1935).

The species can be a pest to lettuce in some areas. The roots may be white with wax produced by the aphids.

39. ***Pemphigus fuscicornis*** (Koch, 1857)

Amycla fuscicornis Koch, 1857: 303. – Survey: 342.

Apterous viviparous female on secondary host (virginogenia). Similar to *bursarius,* but greyish green, not whitish yellow. Abdomen with 2 spinal and 2 pleural wax gland plates on each of the segm. III–VI and 2 plates on VII. Antenna 5-segmented, 0.18 × body; segm. V longer than III+IV; processus terminalis about 0.27 × VIa. Apical segm. of rostrum about 0.09–0.11 mm, 0.8 × 2sht. 1.4–2.2 mm.

Alate viviparous female from secondary host (sexupara). Ant. segm. III frequently shorter than IV+V; VI also frequently shorter IV+V.

Distribution. In Denmark apparently not rare, found in EJ, NWJ, F, and NEZ; in Sweden known from Sk.; not in Norway or Finland. – Distribution in Europe is little known. Originally described and also later recorded from Germany. Known from Poland (incl. the Baltic region) and S Russia.

Biology. The species lives on roots of *Matricaria* (incl. *Tripleurospermum),* maybe also *Chenopodium* and *Beta.* Börner found that *Lapsana* and *Lactuca,* which are secondary hosts of *bursarius,* were not accepted by *fuscicornis.* In Scandinavia the aphids are found exclusively on *Matricaria inodora.* The primary host is unknown.

Pemphigus gairi Stroyan, 1964
Fig. 203.

Pemphigus gairi Stroyan, 1964c: 96. – Survey: 342.

Fundatrix. Pale green. Spinal wax gland plates in some cases absent from pronotum. Otherwise very similar to *bursarius.* 1.9–2.5 mm.

Alate viviparous female from primary host (fundatrigenia). Abdomen yellowish green, in nymphs pale green. Ant. segm. III rather short, 1.5–1.9 × IV and 0.7–1.0 × IV+V; primary rhinarium on penultimate segm. rather large, but rarely broader than 2 × basal diameter of segment (Fig. 203); secondary rhinaria on III: 7–12, IV: 3–6, V: 1–4,

VIa: 1–7. Apical segm. of rostrum and most other characters as in fundatrigenia of *bursarius,* but antennal hairs sometimes longer, on segm. V 0.8–1.7 × basal diameter of segm. (0.5–1.4 in *bursarius,* 0.4–1.0 in *phenax*). 1.5–2.2 mm.

Apterous viviparous female on secondary hosts (virginogenia). Pink or yellowish. Apical segm. of rostrum short, blunt, and rounded apically as in *phenax.* Not to be separated from *phenax* on morphological characters. 1.3–2.2 mm.

Alate viviparous female from secondary host (sexupara). Abdomen dull yellowish. Secondary rhinaria on ant. segm. III: 4–7, IV: 2–4. Apical segm. of rostrum 0.08–0.09 mm, straight-sided. Very similar to *bursarius.* 1.4–2.0 mm.

Distribution. Not yet found in Scandinavia. Stroyan (1964) suggests that Tullgren probably found this species in Stockholm (Upl.) in 1906 (Tullgren 1909: 125; galls of *populinigrae*-type produced on poplars infested by *bursarius*-like sexuparae the year before). – Since *P. gairi* was recognized as a separate species it has appeared to be the most generally abundant of the species forming midrib galls on *Populus nigra* in England. I have not seen published records from other countries, but it is probably a widespread species in Europe.

Biology. The species is host-alternating between *Populus nigra* (I) and roots of *Aethusa cynapium* (II). The gall on *Populus* is similar to that produced by *phenax,* but in an average a little smaller.

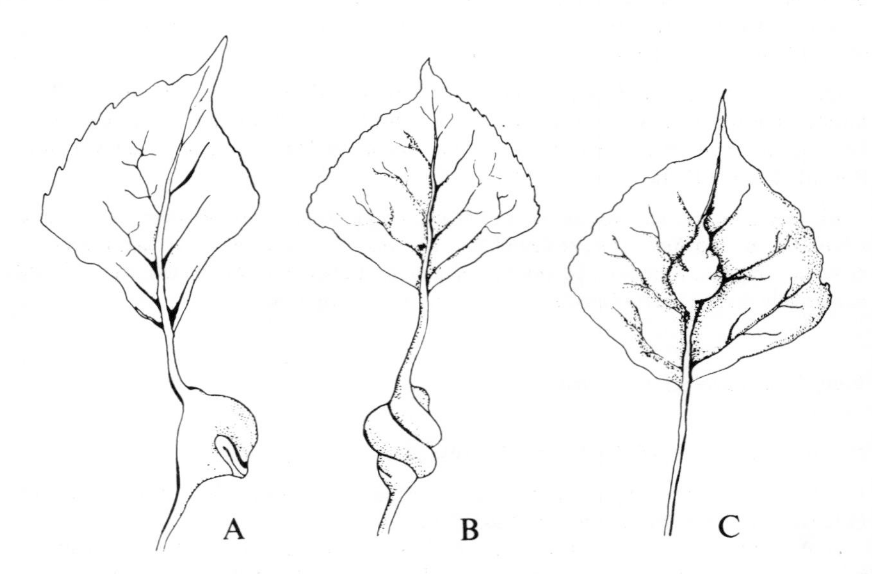

Fig. 200. Galls of some *Pemphigus* spp. on *Populus.* – A: *bursarius* (L.); B: *spyrothecae* Pass.; C: *populinigrae* (Schr.). (After Dixon).

40. ***Pemphigus phenax*** Börner & Blunck, 1916
Figs. 201, 204.

Pemphigus phenax Börner & Blunck, 1916: 31, 39. – Survey: 343.

Fundatrix. Pale green. Very similar to *populinigrae*, but red as a young nymph. 1.9–2.8 mm.

Alate viviparous female from primary host (fundatrigenia). Abdomen yellowish green, in nymphs pale green or bluish green. Primary rhinarium on penultimate ant. segm. usually very broad, rarely narrower than 2 × basal diameter of segment, sometimes covering almost its half length (Fig. 201); secondary rhinaria on ant. segm. III: 6–12, IV: 2–3, V: 0–3, VI: 2–5. Otherwise similar to *bursarius*. 1.5–2.4 mm.

Apterous viviparous female on secondary host (virginogenia). Pale yellow or yellowish white. Very similar to *bursarius*.

Alate viviparous female from secondary host (sexupara). Abdomen dull yellowish. Ant. segm. III shorter than IV+V; secondary rhinaria on III: 4–8, IV: 2–4. Very similar to *bursarius*. 1.9–2.5 mm.

Distribution. In Denmark found on carrots in the Lammefjord (NWZ); in Sweden known from Bl.; not in Norway or Finland. – Europe and W Siberia; common on carrots in England, but apparently rare on *Populus* according to Stroyan (1964); in Germany apparently rare or local, not in N Germany; apparently not rare in Poland.

Biology. The species is host-alternating between *Populus italica* (I) and roots of *Daucus carota*, carrots (II). The gall on *Populus* is an elongate, usually reddish midrib-gall similar to that produced by *populinigrae*, but rather oblong, spindle-shaped, and a

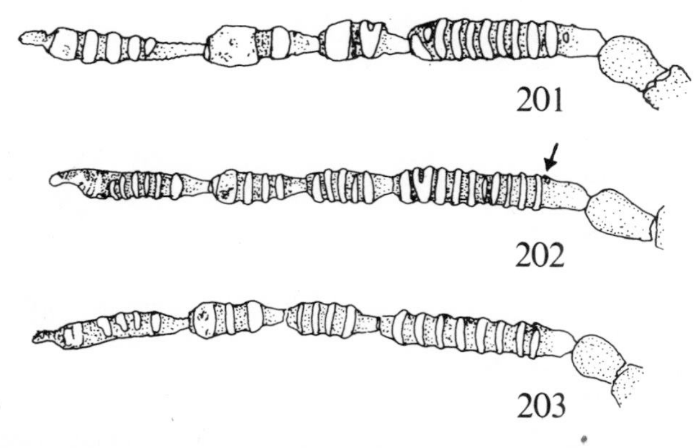

Figs. 201–203. Antennae of alate fundatrigeniae of *Pemphigus* spp. – 201: *phenax* Börn. & Bl.; 202: *protospirae* Licht. (the arrow shows the tooth-like process at base of ant. segm. III); 203: *gairi* Stroyan. (Drawn after photos in Stroyan).

little wrinkled, more or less yellowish laterally (Fig. 204). The alate fundatrigeniae leave the gall through an opening on the underside of the leaf. The galls have not yet been observed in Scandinavia, but the species seems to be a local pest to carrots. Anholocyclic overwintering takes place on carrots e.g. in England, where in some years the infestation is so heavy that the soil appears white with wax wool when the crop is lifted (Stroyan 1964).

41. ***Pemphigus populi*** Courchet, 1879
Figs. 205, 207 A, 208.

Pemphigus populi Courchet, 1879: 90. – Survey: 343.

Fundatrix. Abdominal wax gland plates consisting of rather few relatively large facets (Fig. 207 A). Antenna 0.17 × body; segm. III 0.7–0.95 × IV. Otherwise similar to *bursarius*. 2.4–2.8 mm.

Alate viviparous female from primary host (fundatrigenia). Abdomen with distinct marginal wax gland plates and – on at least some of the posterior tergites – with somewhat larger spinal plates. Ant. segm. III 1.9–2.4 × IV; V about as long as IV, about 0.5 × VI; processus terminalis about 0.25–0.40 × VIa; secondary rhinaria transverse oval, rather broad, on III: 4–6, IV: 2–3, V: 0–2, VI: 0; III without rhinaria on the basal third proximal to the small tooth-like process (Fig. 208). Apical segm. of rostrum pointed, about 0.10–0.12 mm, 0.5–0.7 × 2sht. Siphuncular pores absent (unlike the fundatrigeniae of other *P.* ssp.). 1.4–2.5 mm.

Apterous viviparous female on secondary host (virginogenia). Pale yellow green, covered with wax. Wax gland plates present on abd. segm. (IV–)V–VII. Apical segm. of rostrum 0.11–0.15 mm, 0.8–0.9 × 2sht. Otherwise rather similar to *bursarius*. 1.9–2.6 mm.

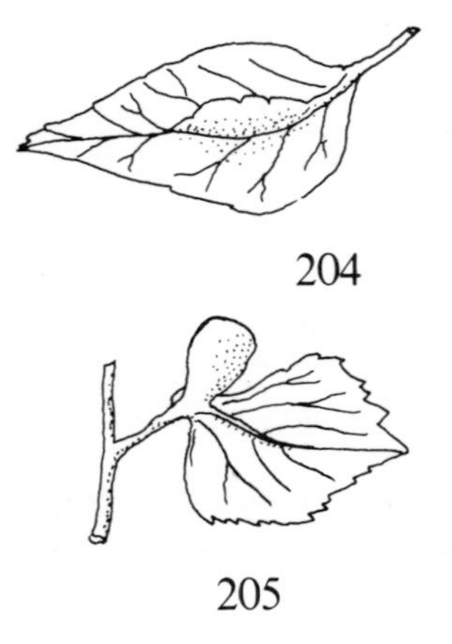

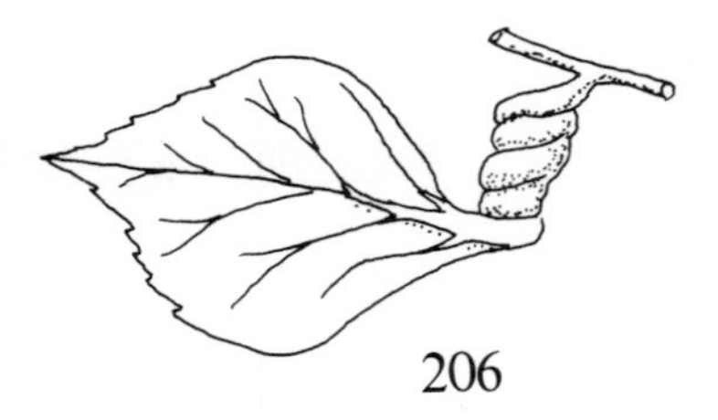

Figs. 204–206. Galls of some *Pemphigus* spp. on *Populus*. – 204: *phenax* Börn. & Bl.; 205: *populi* Courchet; 206: *protospirae* Licht. (After Danielsson).

Alate viviparous female from secondary host (sexupara). Abdomen dark green. Spinal wax gland plates present on abd. segm. I–III(–IV) and a large transverse oval plate on VIII; small marginal plates on I–VII. Ant. segm. III about as long as or shorter than IV+V; VI 1.7–2.1 × V, 0.95–1.1 × IV+V; secondary rhinaria on III: 4–7, IV: 1–2. Apical segm. of rostrum 0.12–0.14 mm, about 0.6 × 2sht., pointed, with 0(–1) accessory hairs. 1.8–2.7 mm.

Distribution. In Sweden in Sm., Ög., and Vg.; not in Denmark, Norway, or Finland. – Widespread in Europe, but apparently rare: British Isles, France, Hungary, Italy, Portugal, Spain; Börner (1952): "Süd- und Mitteleuropa, wenig beachtet"; not recorded from N Germany or Poland; recorded from the USSR (including W Siberia and C Asia) and the Middle East (Turkey, Lebanon).

Biology. The species is host-alternating between *Populus nigra, P. italica,* and *P. balsamifera* (I) and *Melilotus altissimus, Lathyrus pratensis,* and *Medicago lupulina* (II). The gall on *Populus* is oval, more or less irregularly blister-shaped, sometimes almost globular, rather narrow at base, placed on the midrib near the base of the leaf (Fig. 205). The alate fundatrigeniae leave the gall through an opening in the upper end.

42. ***Pemphigus populinigrae*** (Schrank, 1801)
Plate 2: 4. Figs. 200 C, 207 B, 209–211.

Aphis populinigrae Schrank, 1801: 113.
Aphis filaginis Boyer de Fonscolombe, 1841: 188.
Survey: 343.

Fundatrix. Dark green or greyish green, also as a young nymph. Antenna about 0.17 × body; segm. III 1.3–1.5 × IV, sometimes with indication of a subdivision into two segments. Otherwise very similar to *bursarius.* 2.6–2.8 mm.

Alate viviparous female on primary host (fundatrigenia). Abdomen greyish green or dark green, with 6 longitudinal rows of more or less fused wax gland plates. Ant. segm. III about 1.3 × IV+V; processus terminalis not well defined, rather broad at base (Fig. 209); secondary rhinaria rather narrow, on III: 12–18, IV: 3–7, V: 4–7, VI: 4–7. Apical segm. of rostrum rather long, 0.10–0.12 mm, about 0.6 × 2sht., straight-sided. Wing veins not shadowed. Other characters as in *bursarius.* 1.8–2.9 mm.

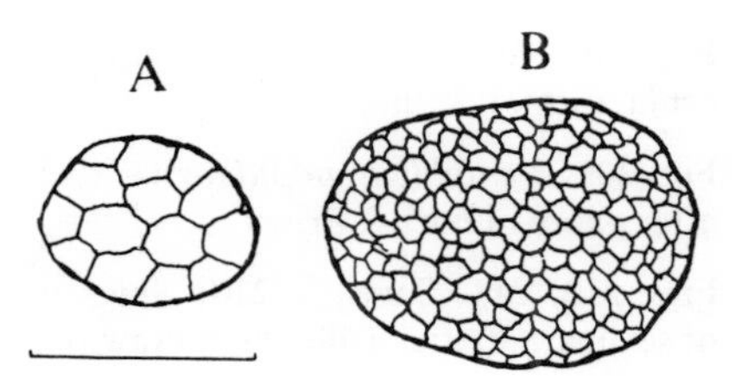

Fig. 207. Wax gland plates of fundatrices of A: *Pemphigus populi* Courchet, spinal plate on abd. segm. II; B: *P. populinigrae* (Schr.), pleural plate on metathorax (one of the smaller plates of this species). (Scale 0.05 mm).

Apterous viviparous female on secondary host (virginogenia). Yellowish or green, strongly covered with wax wool. Abd. tergites III–VI each with 4 wax gland plates, VII with 2 plates. Processus terminalis 0.17–0.20 × VIa (Va). Apical segm. of rostrum long and pointed, about 0.15 mm, 1.2–1.4 × 2sht. Otherwise similar to *bursarius*. 2.0–2.2 mm.

Alate viviparous female from secondary host (sexupara). Very similar to *bursarius*, but wax gland plates apparently absent from thorax, and apical segm. of rostrum long and pointed as in the virginogenia described above. Secondary rhinaria on ant. segm. III: 7–9, IV: 2–4 (Fig. 211). 1.7–2.5 mm.

Distribution. In Denmark common and widespread; in Sweden common and widespread north to Upl.; in Norway found in VAy; in Finland north to Oa and Sb. – In Europe from the British Isles and France east to Poland and the USSR, south to Portugal, Spain, and Bulgaria; common in N Germany, also at the Baltic coast of Poland. N Africa and W Asia (Turkey, Lebanon). W Siberia and C Asia.

Biology. The species is host-alternating between *Populus nigra, P. italica, P. balsamifera, P. canadensis* a. o. *P.* ssp. (I) and *Gnaphalium* and *Filago* (II). The gall on

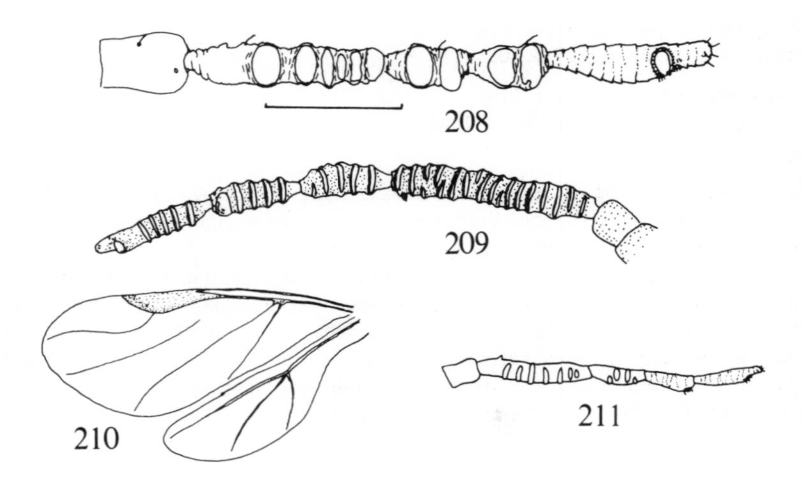

Fig. 208. *Pemphigus populi* Courchet, antennal segments II–VI of alate fundatrigenia. (Scale 0.1 mm).

Fig. 209. *Pemphigus populinigrae* (Schr.), antenna of alate fundatrigenia. (Drawn after photo in Furk & Prior).

Figs. 210, 211. Same. – 210: wings of alate fundatrigenia: 211: ant. segm. II–VI of sexupara. (After Tullgren, redrawn).

Populus is an elongate, rather broad, usually bright red, blister-shaped swelling, about 10 mm long and 5–7 mm broad. It is placed on the midrib on the upperside of the leaf, usually not very close to the leaf base (Fig. 200 C). The central part of the gall is roundish or oval and smooth. The alate fundatrigeniae leave the gall through a narrow opening, which appears on the underside of the leaf when the gall has ripened in June–August. The woolly colonies of virginogeniae are found mainly above the ground on the stems and leaves of the secondary hosts. The sexuparae return to *Populus* in September–October to bear the small sexuales, but the colonies may continue reproduction by parthenogenesis on the secondary hosts through winter (Mordvilko 1935).

43. ***Pemphigus protospirae*** Lichtenstein, 1885
Figs. 202, 206.

Pemphigus protospirae Lichtenstein, 1885: 31. – Survey: 344.

Fundatrix. Antenna 0.20 × body. Very similar to *bursarius*. About 2.2 mm.

Alate viviparous female from primary host (fundatrigenia). Abdomen dirty greyish green. Wax gland plates absent from the middle part of abdomen. Ant. segm. III a little longer than IV+V; VI about twice as long as IV; processus terminalis well defined and finger-like; secondary rhinaria on III: 10–11, IV: 4–5, V: 3, VIa: 6; no secondary rhinaria placed proximal to the tooth-like process near base of III (Fig. 202). Apical segm. of rostrum rather long, 0.10–0.12 mm, but otherwise similar to *bursarius*.

Apterous viviparous female on secondary host (virginogenia). Yellowish or greenish. As *bursarius*, but the apical segm. of rostrum sometimes a little longer, 0.10–0.13 mm, 0.8–1.0 × 2sht. 1.6–2.3 mm.

Alate viviparous female from secondary host (sexupara). Abdomen yellowish green. Secondary rhinaria on ant. segm. III: 4–9, IV: 1–4. Very similar to *bursarius*, but apical segm. of rostrum with 0–2 accessory hairs. 1.5–2.3 mm.

Distribution. In Sweden found in Sk. and Öl.; not in Denmark, Norway, or Finland, but it has possibly been overlooked because the gall is similar to that of the very common *P. spyrothecae*. – Widespread in Europe, including the British Isles, N Germany, the Baltic region of Poland, and W Russia, but more common in S Europe than in N & C Europe. Outside Europe recorded from W Siberia and C Asia. Records from Turkey and Iran are doubtful.

Biology. The gall on *Populus italica* is similar to that of *P. spyrothecae*. The petiole is flattened and spirally twisted (Fig. 206). The size and shape of the gall are variable. Sometimes the gall consists of the whole petiole and part of the leaf base as well. It is smooth, shiny, green, or green mottled with red. The alate fundatrigeniae leave the gall in July in Sweden (in France as early as in May–June) (Tullgren 1909). The secondary hosts are *Berula erecta, Sium latifolium,* and *Apium nodiflorum*. The virginogeniae inhabit the sheathing leaf bases of plants growing in water and the roots of plants growing on the bank.

44. ***Pemphigus saliciradicis*** (Börner, 1950)
Figs. 212–214.

Parathecabius saliciradicis Börner, 1950: 18.
Pemphigus salicicola Hille Ris Lambers, 1952: 31.
Survey: 344.

Apterous viviparous female (virginogenia). Head often with 2 small wax gland plates (observed in a specimen from Sweden, also found by Aoki (1975) in some specimens from Japan). Abdomen with 4 large plates on each of the segm. III–VI, and 2 plates on VII. Antenna 5- or 6-segmented, 0.25–0.33 × body; segm. VI longer than IV+V (in 6-segmented antenna) in specimens from Sweden and Greenland, shorter than IV+V in specimens from Japan; processus terminalis well-defined, from about 0.2 to 0.33 × VIa. Apical segm. of rostrum about 0.09–0.10 mm, almost parallel-sided, with abruptly rounded, very blunt apex (Fig. 212). Hind tarsus with numerous spinules on distal margin of first tarsal segm. and the sole of second tarsal segm. (Fig. 213). 1.6–2.6 mm.

Alate viviparous female (virgino-sexupara). Wax gland plates absent from head as in alatae of other *P.* spp.; marginal and several more or less well developed dorsal wax gland plates present on abdomen. Antenna (Fig. 214) about 0.25 × body; segm. III shorter than IV+V; primary rhinaria surrounded by short hairs; secondary rhinaria on III: 6–8 on distal 2/5, IV: 2–4, V: 0–1, VI: 0. Apical segm. of rostrum 0.10 mm, about 0.55 × 2sht., blunt, with 0–3 short accessory hairs. 2.1–2.4 mm.

Distribution. In Sweden found by Danielsson (unpublished) in T. Lpm.; not in Denmark or Norway; in Finland found in Ta (Janakkala) by Heikinheimo. – The British Isles, Switzerland; Greenland (Egedesminde and Kapisigdlit); Japan (near Sapporo).

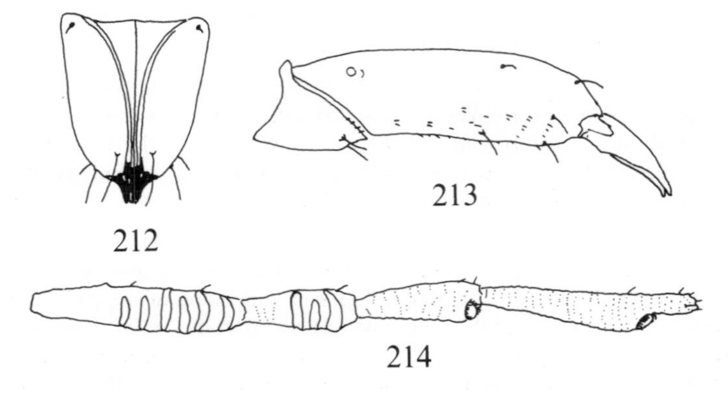

Figs. 212–214. *Pemphigus saliciradicis* (Börn.). – 212: apical segment of rostrum of apt. viv.; 213: hind tarsus of apt. viv.; 214: ant. segm. III–VI of al. viv. (212, 213 after Furk & Prior, 214 after Aoki; redrawn).

Biology. The species has been found only on the roots of *Salix* spp. *(polaris, herbacea, fragilis, alba,* perhaps also *lapponum).* Alatae have been found only in Japan (Hokkaido). They are similar to sexuparae produced by colonies on the secondary hosts of host-alternating *Pemphigus* spp. with regard to antennal characters, but Aoki (1975) found that some of them only contain rostrate embryos, while others contain both rostrate and arostrate embryos. True sexuparae containing arostrate embryos, only, were not observed.

45. ***Pemphigus spyrothecae*** Passerini, 1856
Plate 1: 8 and 2: 3. Figs. 42, 200 B.

Pemphigus spyrothecae Passerini, 1856.
Pemphigus spirothecae Passerini, 1860: 39.
Survey: 345.

Fundatrix. Light green. Ant. segm. III about 0.8 × IV. Otherwise very similar to *bursarius.* 1.9–2.0 mm.

Apterous viviparus female (apterous fundatrigenia). Antenna 6-segmented, about 0.25 × body; segm. VI longer than IV+V. Apical segm. of rostrum short, blunt, with slightly convex margins, about 0.09 mm, 0.6 × 2sht. Siphuncular pores present. About 1.7 mm.

Alate viviparous female (sexupara). Abdomen green. Wax powdered. Wax gland plates present on thorax and abdomen; pronotum with 4 plates; mesonotum usually with 2 small spinal plates; metanotum also with 2 plates; abd. segments with marginal plates and better developed spinal plates. Ant. segm. III as long as IV+V, VI a little shorter; secondary rhinaria on III: 5–6, on distal half, IV: 1–4. Apical segm. of rostrum with 0–2 accessory hairs as in sexupara of *protospirae,* but the segment shorter, 0.09 mm, about 0.5 × 2sht. Most other characters as in sexupara of *bursarius.* 1.9–2.2 mm.

Sexuales. Antenna 4-segmented. Very small; oviparous female (Fig. 42) about 0.8 mm.

Distribution. In Denmark very common and widespread; in the southern part of Sweden very common, north to Upl.; in the southern part of Norway, north to TEi; not in Finland. – Common and widespread in Europe, including Great Britain, N Germany, the Baltic region of Poland, and NW & W Russia, south to Portugal, Spain, and Bulgaria. N Africa: Tunesia. Records from the Middle East are doubtful. W Siberia. Introduced in Canada.

Biology. The species is not host-alternating. The holocycle consists of three or four generations on *Populus nigra, P. italica,* or *P. balsamifera.* The gall is formed on the petiole, which is flattened and spirally twisted (Fig. 200 B). It is green, red, or yellowish red, and smooth. The fundatrix-generation is followed by a generation of apterous viviparous females (apterous fundatrigeniae). Alate viviparous females develop at the end of the summer and in autumn. Some of them are daughters of the apterous fundatrigeniae and so belong to the third generation, while others are the last-born daughters of the fundatrix and so belong to the second generation. All alatae are sexuparae, which leave the galls in autumn to bear sexuales on the bark of trunks and branches. The galls may still contain live aphids as late as in November.

Pemphigus trehernei Foster, 1975
Figs. 215–217.

Pemphigus trehernei Foster, 1975: 257. – Survey: 345.

Fundatrix. Dull pale green. Wax gland plates as in *bursarius;* spinal plates on pronotum adjoined by small, scattered, pigmented spots, but without well-defined pigmented area; marginal plates on pronotum not enclosed on posterior margin by pigmented area. Antenna about 0.10 × body; segm. III a little longer than IV. Apical segm. of rostrum 0.09 mm, 0.65 × 2sht. 2.9–3.0 mm.

Alate viviparous female from primary host (fundatrigenia). Abdomen dull pale green. Wax gland plates as in *bursarius*. Antenna (Fig. 215) about 0.4 × body; segm. III 2.2–2.6 × IV, 0.9–1.1 × IV+V; processus terminalis 0.14–0.15 × VIa; VI shorter than IV+V; primary rhinarium on V round, not large, nor inflated; secondary rhinaria on III: 11–19, IV: 4–8, V: 4–10, VIa: 6–11. Apical segm. of rostrum 0.08–0.10 mm, 0.4–0.5 × 2sht., with slightly convex margins. Wing veins slightly shadowed with brownish. 1.9–2.4 mm.

Apterous viviparous female on secondary host (virginogenia). Yellowish white or greenish white, with greyish wax powder, on posterior part of abdomen with tufts of wax. Wax gland plates as in *bursarius*. Antenna 5-segmented, rarely 6-segmented; processus terminalis 0.3–0.4 × Va; V in 5-segmented antenna 0.09–0.15 mm, as long as or longer than III+IV. Apical segm. of rostrum 0.07–0.11 mm, 0.5–0.6 × 2sht., almost parallel-sided, blunt. 1.3–2.4 mm.

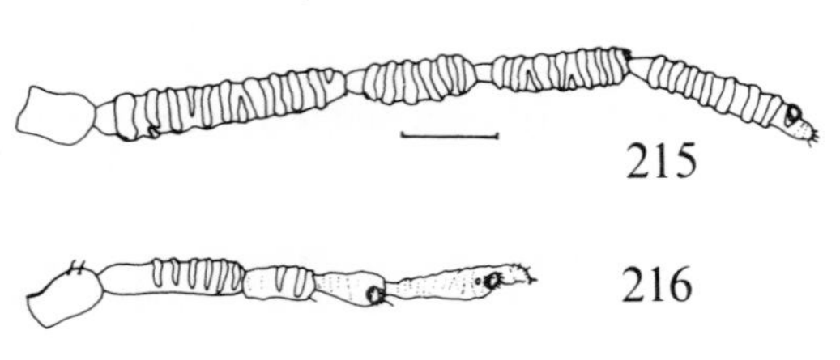

Figs. 215, 216. *Pemphigus trehernei* Foster, antennal segments II–VI of al. viv. – 215: fundatrigenia. (Scale 0.1 mm); 216: sexupara. (After Foster, redrawn).

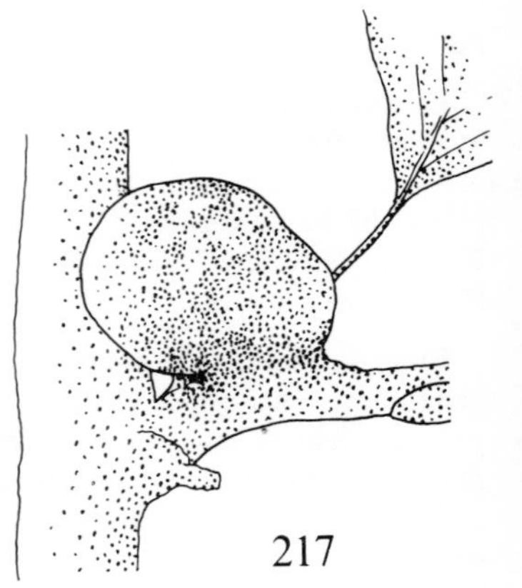

Fig. 217. Same, gall on *Populus*. (Drawn after photo in Foster).

Alate viviparous female from secondary host (sexupara). Abdomen yellow or – in old individuals – greenish yellow. With tufts of wax on thorax and abdomen. Wax gland plates as in *bursarius*. Antenna (Fig. 216) 6-segmented, about 0.25–0.3 × body; segm. III 1.0–1.7 × IV+V; VI 1.0–1.4 × IV+V, 2.0–2.8 × V; processus terminalis about 0.2 × VIa; secondary rhinaria on III: 3–7, IV: 1–3, placed on distal 2/3 of the segments. Apical segm. of rostrum 0.07–0.09 mm, 0.5–0.65 × 2sht., almost parallel-sided, without accessory hairs. 1.4–2.1 mm.

Distribution. Not yet found in Scandinavia, but possibly overlooked. – Recently described from the British Isles, here found in tidal salt-marshes in England and Ireland; also recorded from the Gironde-region in France by Foster (1975).

Biology. The species is host-alternating between *Populus italica* (I) and *Aster tripolium* (II), but the majority of the aphids overwinter on the roots of the latter in salt-marsh habitats. The species has only been found on the secondary host, but Foster transferred sexuparae to *Populus italica,* which was accepted as a primary host. Two eggs survived the winter, and two galls were produced next year in June. Alate fundatrigeniae from these galls were allowed access to *Aster tripolium* and they produced apterous virginogeniae on this plant. The gall is green with red tinge around aperture, globular, with maximum diameter about 1.8 cm, rather similar to the gall of *bursarius,* but formed on the base of the current year's shoot, not on the petiole (Fig. 217).

SUBFAMILY FORDINAE

Antennae of alate females 5- or 6-segmented; secondary rhinaria circular, subcircular, or transverse oval, in *Melaphis* narrow and ringlike. Media of fore wing usually unbranched, in *Melaphis* with one fork. Hind wing usually with two oblique veins. First tarsal segm. with 4 or more hairs, or tarsi one-segmented (in aptera of *Melaphis*). Siphuncular pores absent from all morphs. Facetted wax gland plates present or absent.

All Scandinavian species are anholocyclic. They reproduce exclusively by parthenogenesis on roots of herbs, mostly grasses, or on mosses. In warmer climates some of these species or close relatives are holocyclic and host-alternating, using members of the plant family Anacardiaceae *(Pistacia* or *Rhus)* as the primary hosts and herbs, especially grasses or mosses as the secondary hosts. The holocycle differs from that of Eriosomatinae, Pemphiginae and most other Aphidoidea in being biennial. Overwintering as eggs takes place on the primary hosts, and overwintering as parthenogenetic females takes place on the secondary hosts. The colonies founded by the fundatrigeniae on the secondary hosts will not produce sexuparae until the following year. Continued reproduction by parthenogenesis in colonies of virginoparae has been observed not only in cold climates, where the primary hosts do not grow, but also in areas where holocyclic populations occur.

The subfamily is rather heterogeneous. Baker (1920) placed genera with well developed wax gland plates in the tribe Melaphini (including *Melaphis* and *Aploneura*) and genera with reduced wax gland plates in the tribe Fordini *(Forda, Geoica, Paracletus).* Börner (1930) subdivided Fordinae into two groups, viz. A: genera with *Pistacia* as the primary host of holocyclic species (*Paracletus, Forda, Geoica, Aploneura* a. o.), and B: genera with *Rhus* as the primary host of holocyclic species (*Melaphis* a. o.), but could not give morphological differences between the gall producers on *Pistacia* and *Rhus,* and wrote: "Ob solche bestehen oder ob die an beiden Gattungen der Anacardiaceen lebenden Arten keine entsprechenden Verwandtschaftsgruppen bilden, bedarf der Prüfung". The former group was later subdivided by Börner (1952) into two tribes, Fordini without hairs surrounding the primary rhinaria *(Paracletus, Forda, Smynthurodes),* and Baizongiini with such hairs present, subdivided into two subtribes, Baizongiina with facetted wax gland plates *(Baizongia, Aploneura)* and Geoicina without such plates *(Geoica).* Zwölfer (1958) regarded these groups as tribes and changed the sequence in accordance with his view that the presence of wax gland plates and haired rhinaria are original (plesiomorphous) characters, so that Fordini shall be regarded as the tribe diverging most from the common ancestor through reduction.

Roundish secondary rhinaria of alatae are mentioned as a diagnostic character of Fordinae by Börner (1952) and also Zwölfer (1958), who both treated solely the Central European genera. The nearctic *Melaphis* has ringlike secondary rhinaria of alatae and is different from the European Fordinae in several other respects, so it is placed in a tribe of its own in the present paper, and the tribes of Zwölfer are regarded as subtribes of another tribe, Fordini. I am not sure if some genera in southern and eastern Asia shall be placed in the same tribe as *Melaphis.*

It is hard to say if ringlike rhinaria is a plesiomorphous character or an apomorphous one, in this group of aphids probably plesiomorphous because it is also found in other Pemphigidae. *Melaphis* has some autapomorphies, however, e.g. one-segmented tarsi of apterae.

Two keys are given below to the Fordinae of Scandinavia, one to tribes and subtribes and one to genera.

Key to tribes and subtribes of Fordinae

1 Secondary rhinaria of alatae ringlike. Apterae with one-segmented tarsi .... **Melaphidini** (p. 206)
– Secondary rhinaria of alatae circular, subcircular, or oval. Apterae with two-segmented tarsi. .... **Fordini** – 2
2 (1) Primary rhinaria not surrounded by short hairs. .... **Fordina** (p. 199)
Primary rhinaria surrounded by short hairs. 3
3 (2) Wax gland plates of apterae well developed. .... **Baizongiina** (p. 192)
– Wax gland plates reduced. .... **Geoicina** (p. 195)

Key to genera of Fordinae

Apterous viviparous females

1 Tarsi one-segmented (Fig. 246). On moss. .... *Melaphis* Walsh (p. 206)
– Tarsi two-segmented. Not on moss. 2
2 (1) Primary rhinaria surrounded by short hairs (Fig. 226). 3
– Primary rhinaria not surrounded by hairs. 5
3 (2) Dorsum without wax gland plates; with spatulate (or flabellate) hairs (Fig. 225 B). .... *Geoica* Hart (p. 196)
– Dorsum with six longitudinal rows of wax gland plates (Figs. 218, 221); without spatulate hairs. 4
4 (3) Body elliptical or spindle-shaped. Abd. tergite VII with 4 wax gland plates. .... *Aploneura* Passerini (p. 192)
– Body broad, oval or almost globular. Abd. tergite VII with 2 wax gland plates. .... *Baizongia* Rondani (p. 194)
5 (3) Antenna 6-segmented. Cuticle reticulate. Body flattened. *Paracletus* v. Heyden (p. 199)
– Antenna 5-segmented. Cuticle not reticulate, or reticulate only on head and prothorax. Body not flattened. 6
6 (5) Primary rhinaria with thick, sclerotized rims (Fig. 242). Ant. segm. III sometimes shorter than II, never twice as long as II (Fig. 241). .... *Smynthurodes* Westwood (p. 204)
– Primary rhinaria without thick, sclerotized rims. Ant. segm. III about twice as long as II or longer (Figs. 235, 236). ... *Forda* v. Heyden (p. 201)

Alate viviparous females

1 Media of fore wing with one fork (Fig. 245). Secondary rhinaria narrow, ringlike (Fig. 247). .... *Melaphis* Walsh (p. 206)
– Media of fore wing unbranched. Secondary rhinaria circular or transverse oval. 2
2 (1) Ant. segm. III and IV each with a secondary rhinarium surrounded by short hairs, placed in the distal end of the segment as the primary rhinaria on segm. V and VI (Fig. 220). Cubital veins of fore wing with a common basal stem (Fig. 219). Wings flat in repose. .... *Aploneura* Passerini (p. 192)
– Ant. segm. III with several secondary rhinaria. Cubital veins of fore wing sometimes leaving the main vein from the same point, but without a common basal stem (Fig. 238). Wings rooflike in repose. 3
3 (2) Primary rhinaria surrounded by short hairs. 4
– Primary rhinaria not surrounded by hairs. 5

4 (3) With spatulate body hairs. .... *Geoica* Hart (p. 196)
– Without spatulate body hairs. .... *Baizongia* Rondani (p. 194)
5 (3) Antenna 6-segmented. Ant. segm. IV with more than 10 secondary rhinaria. .... *Paracletus* v. Heyden (p. 199)
– Antenna 5- or 6-segmented. Ant. segm. IV with less than 10 secondary rhinaria. 6
6 (5) Primary rhinaria with thick, sclerotized rims. ... *Smynthurodes* Westwood (p. 204)
– Primary rhinaria without thick, sclerotized rims. .... *Forda* v. Heyden (p. 201)

TRIBE FORDINI

Apterae with two tarsal segments. Secondary rhinaria of alatae roundish. Media of fore wing unbranched. Fundatrix and fundatrigeniae of holocyclic species (S Europe) in galls on *Pistacia.*

With three subtribes in Europe.

SUBTRIBE BAIZONGIINA

Primary rhinaria surrounded by short hairs. Apterae with well developed spinal, pleural, and marginal wax gland plates.

With two genera in Scandinavia.

Genus *Aploneura* Passerini, 1863

Aploneura Passerini, 1863: 201.

Type-species: *Tetraneura lentisci* Passerini, 1856.

Survey: 92.

Apterae with 4- or 5-segmented antennae, alatae with 6-segmented antennae. Both apterous and alate virginogeniae with facetted wax gland plates, two rudimentary gonapophyses, and primary rhinaria surrounded by short hairs. Ant. segm. III and IV of alatae each with one secondary rhinarium similar to the primary rhinaria on V and VI, placed in the distal ends of the segments (Fig. 220). Wings flat in repose; fore wing with bases of cubital branches united (Fig. 219); hind wing with one oblique vein or with two oblique veins, the proximal one then being more or less obliterated.

With 3 species in the world, one species in Scandinavia.

46. ***Aploneura lentisci*** (Passerini, 1856)
Figs. 218–220.

Tetraneura lentisci Passerini, 1856: 264.
Rhizobius graminis Buckton, 1883: pl. 129.
Survey: 92.

Apterous viviparous female. Pale yellowish. Prothorax, antennae, legs, and anal region brownish; covered with wax. Body elliptical or spindle-shaped. Wax gland plates present on head, thorax, and abdomen; one spinal, one pleural, and one marginal pair of plates present on most segments (Fig. 218); head with 2 spinal pairs; abd. tergite VII with one spinal and one marginal pair. Eyes consisting of three ommatidia. Antenna about 0.11 × body; segm. II in 5-segmented antenna about as long as III or longer; V longer than any other segm.; processus terminalis about 0.33 × Va. Rostrum reaching middle coxae. Trochantera and femora fused. 1.5–2.4 mm.

Alate viviparous female. Abdomen green or yellowish green, with wax secretions. Ant. segm. III as long as IV and about as long as or somewhat shorter than V, about 0.5 × VI (Fig. 220); processus terminalis about 0.2 × VIa. Rhinaria and wings as described for the genus. 1.7–2.0 mm.

Distribution. In Denmark found at Skive (NWJ); not in Sweden, Norway, or Finland. – The origin of this species is probably the Mediterranean region, where the primary host is native and common; now widespread in Europe, Asia, Africa, and Australia; also recorded from N Zealand and Argentina.

Biology. The species is holocyclic in the Mediterranean region, producing leaf galls on the primary host, *Pistacia lentiscus*. The alate fundatrigenia, which flies to grass roots, is described and depicted by Mordvilko (1935). It is similar to the alate virginogenia described above. In N & C Europe and C Asia the species is exclusively anholocyclic, feeding on grass roots all the year round. The secondary hosts, in N & C Europe the only hosts, are various grasses, e.g. *Poa, Triticum, Agropyrum, Bromus, Dac-*

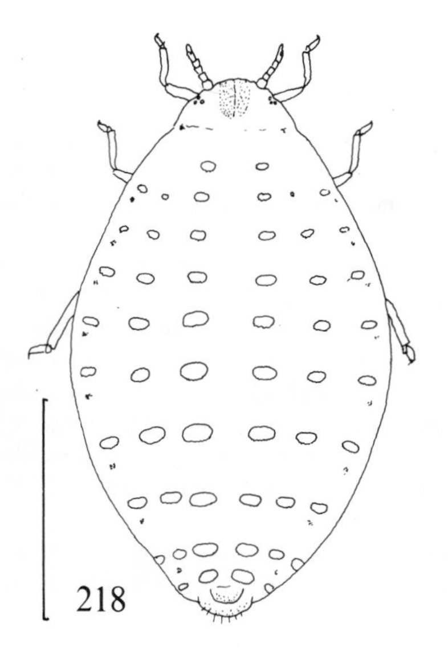

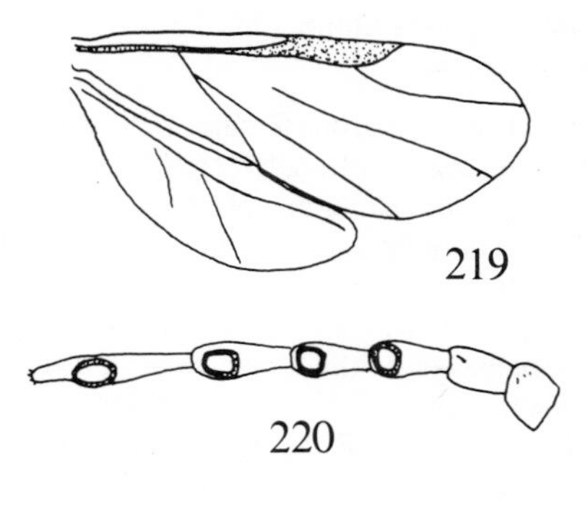

Fig. 218. *Aploneura lentisci* (Pass.), apt. viv. (Scale 1 mm).

Figs. 219, 220. Same, al. viv. – 219: wings; 220: antenna. (After Mordvilko, redrawn).

tylis, Agrostis, and *Hordeum.* In Denmark it has only been found on roots of *Poa annua* on rather heavy clayish soil. The colonies are easy to observe because the roots look white on account of the considerable amount of wax produced by the aphids. They are not visited by ants.

Genus *Baizongia* Rondani, 1848

Baizongia Rondani, 1848: 35.
 Type-species: *Aphis pistaciae* Linné, 1767.
Pemphigella Tullgren, 1909: 171.
 Type-species: *Tetraneura cornicularia* Passerini, 1856.
 (errore pro *Pemphigus cornicularius* Passerini, 1856). = *Aphis pistaciae* Linné, 1767.
Survey: 104.

Apterae with 5-segmented antennae, alatae with 6-segmented antennae. Both apterous and alate virginogeniae with facetted wax gland plates and two rudimentary gonapophyses. Primary rhinaria surrounded by short hairs, in alatae at least on ant. segm. VI. Ant. segm. IV of alatae with 1–2 secondary rhinaria, one of them (or the only one) placed in the distal end of the segment like the primary rhinarium on V. Wings roof-like in repose contrary to the condition in *Aploneura.*

Two species in the world, one species in Scandinavia.

47. ***Baizongia pistaciae*** (Linné, 1767)
 Figs. 221–223.

Aphis pistaciae Linné, 1767: 737.
Survey: 104.

Apterous viviparous female. Whitish or yellowish, wax powdered. Head, prothorax, antennae, legs, and anal region brownish. Body egg-shaped, almost globular, with distinctly convex margins of pronotum. Head with 3 pairs of spinal wax gland plates (Fig. 222). Thoracic segments and abd. segm. I–VI each with spinal and marginal plates, and also with very small pleural plates except on pronotum; abd. segm. VII only with marginal plates. Eyes with three ommatidia. Antenna 0.17–0.25 × body; ant. segm. III as long as or a little longer than II, usually longer than IV, and distinctly shorter than V, which is about 0.33 × antennal length; processus terminalis about 0.5 × Va; secondary rhinaria absent. Rostrum reaching past middle coxae; apical segm. 1.3–1.4 × 2sht. Anus in dorsal position; two longitudinal rows of long hairs on the prolonged, almost rectangular, anal plate form together with the hairs on abd. tergite VII a so-called trophobiotic organ to keep the honeydew until it can be removed by an ant (Fig. 223). 1.4–2.4 mm.

Alate viviparous female. Ant. segm. III with more than one secondary rhinarium. For other differences between the alatae of *Baizongia* and *Aploneura* see the descriptions of genera.

Distribution. In Sweden known from Sk., Sm., Öl., and Sdm.; not in Denmark, Norway, or Finland. – Widespread in Europe, incl. the British Isles, the Netherlands, Germany (but not in N Germany), and Poland. Over the entire Mediterranean region, incl. N Africa and the Middle East; in Asia also in Pakistan, NW Himalayas.

Biology. The species is anholocyclic on grass-roots in N & C Europe, and holocyclic in S Europe with host-alternation between *Pistacia terebinthus* (I), where large galls similar to cows' horns are produced, and grass-roots (II). Among the grasses used as hosts are *Agrostis, Corynephorus, Dactylis, Festuca,* and *Poa.* It is attended by ants, especially *Lasius flavus,* which may bring the aphids into the nests in autumn. In spring the aphids are again dispersed on grass-roots around the nests. Alatae are extremely rare in N Europe. They have not yet been observed in Scandinavia.

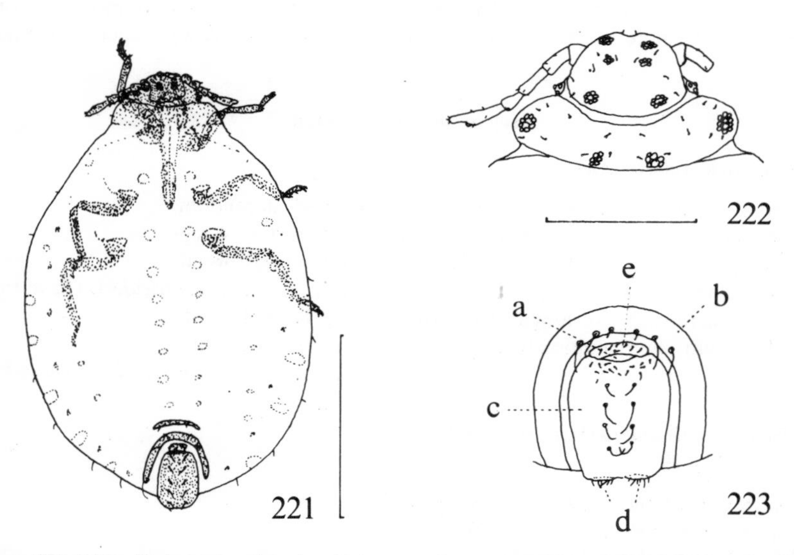

Figs. 221–223. *Baizongia pistaciae* (L.), apt. viv. – 221: body (drawing of transparent mounted specimen; rostrum and part of the legs visible only from the underside in live specimens); 222: head and prothorax in dorsal view, with wax gland plates; 223: anal region (dorsal view); a = anus, b = abd. tergite VIII, c = anal plate, d = rudimentary gonapophyses, e = cauda. (Scales 1 mm for 221, 0.2 mm for 222 & 223). (222 & 223 after Zwölfer, redrawn).

SUBTRIBE GEOICINA

Primary rhinaria surrounded by short hairs. Apterae without wax gland plates.

Genus *Geoica* Hart, 1894

Geoica Hart, 1894: 101.
Type-species: *Geoica squamosa* Hart, 1894.
= *Pemphigus utricularius* Passerini, 1856.
Survey: 203.

Apterae with 4- or 5-segmented antennae, alatae with 6-segmented antennae. Apterous virginogeniae without wax gland plates; cuticle reticulate; anus in dorsal position (Fig. 228); body hairs pointed or spatulate to flabellate (Fig. 225). Both apterae and alatae with two rudimentary gonopophyses and primary rhinaria surrounded by short hairs (Fig. 226).

With 8 species in the world, perhaps more (if *G. utricularia* consists of more than one species, see the note to this species or complex of species below), 2 species in Scandinavia.

Key to species of *Geoica*

Apterous viviparous females

1 Anal plate with numerous short hairs dispersed at random. Primary rhinarium on penultimate ant. segm. (III in 4-segmented antennae, IV in 5-segmented antennae) broadly oval to circular. .... 48. *utricularia* (Passerini)

– Anal plate with two longitudinal rows of long hairs. Primary rhinarium on penultimate ant. segm. transverse oval, narrow. .... 49. *setulosa* (Passerini)

48. ***Geoica utricularia*** (Passerini, 1856)
Figs. 224–229.

Pemphigus utricularius Passerini, 1856: 260.
Tychea eragrostidis Passerini, 1860: 39.
Endeis carnosa Buckton, 1883: 92.
Geoica squamosa Hart sensu Tullgren, 1925: 38.
Survey: 204.

Apterous viviparous female. Dirty brownish white or yellowish white, faintly wax powdered. Head, antennae, legs, and anal region brownish. Body broadly oval. Hairs on body and appendages variable, pointed, spatulate, or flabellate (Fig. 225). Wax gland plates absent, but wax gland pores present in the polygonal fields of the reticulate pattern of the cuticle. Eyes with three ommatidia. Antenna about 0.3 × body; segm. III much longer than the ultimate segm., also in 5-segmented antennae; primary rhinaria subcircular or broadly oval (Fig. 226); secondary rhinaria absent. Rostrum reaching past middle coxae; apical segm. 1.5–1.6 × 2sht. Anus in dorsal position (Figs. 228, 229);

anal plate prolonged, almost rectangular, with numerous short hairs, which are not arranged in rows (Fig. 229). 1.3–2.2 mm.

Alate viviparous female. Abdomen pale green, with dark, dorsal, transverse stripes. Anal region and venter wax powdered. Cuticle without reticulation. Antenna about 0.3 × body; segm. III with 3–9 round or oval secondary rhinaria of different sizes, IV: 1–2, V: 0–1, VI: 0 (Fig. 227). Cubital veins of fore wing leave the main vein from the same point. Oblique veins of hind wing widely separated. About 1.8–1.9 mm.

Distribution. In Denmark widespread (EJ, WJ, NWJ, NEJ, NWZ); in Sweden known from Sk., Sm., Öl., Ög., and Upl. (the records from Ög. and Upl. (Stockholm) refer to Tullgren (1925)); not in Norway; found in Finland (N) – Widespread in Europe, incl. the British Isles, Germany (also in N Germany), Poland (not in the Baltic region), and NW & W Russia, south to Portugal, Spain, and Italy; N Africa; Asia: Turkey, Lebanon, Israel, Syria, Iraq, Iran, W Siberia, C Asia; N America: widespread in the USA.

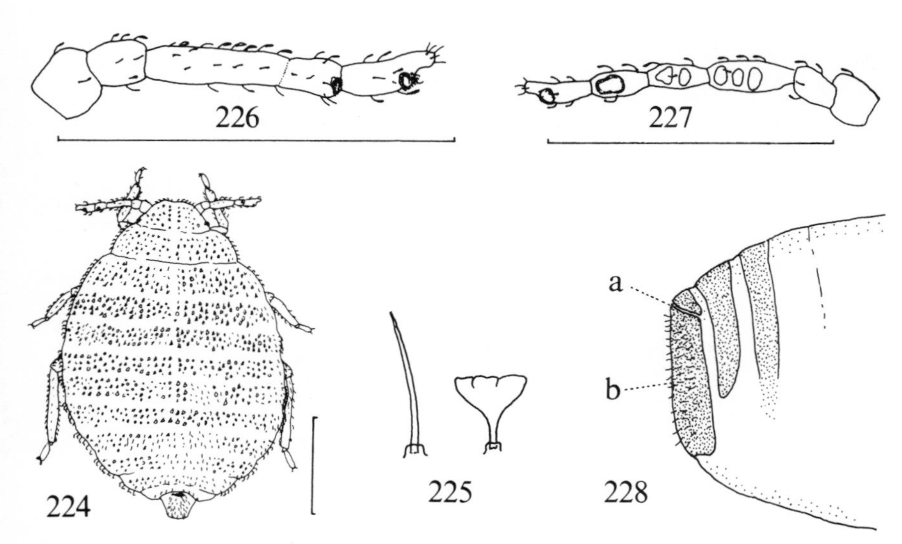

Figs. 224, 225. *Geoica utricularia* (Pass.). – 224: body of apt. viv. in dorsal view: 225: types of hairs, pointed and flabellate. (224 after Mordvilko, redrawn).

Figs. 226, 227. Same. –226: antenna of apt.viv.; 227: antenna of al. viv. (virginopara). (After Mordvilko, redrawn; small hairs surrounding primary rhinarium of ant. segm. V added in accordance with Abb. 16 in Zwölfer).

Fig. 228. Same, posterior end of apt. viv. seen from the right side; a = anus, b = anal plate. (All scales 0.5 mm).

Biology. The species is in N & C Europe anholocyclic on grass-roots and is often located rather deep in the soil. In S Europe it is holocyclic and host-alternating between *Pistacia terebinthus* (I) and grasses (II). The galls on *Pistacia* are large, red and placed on the underside of the midrib of the leaves. Among the hosts in N & C Europe are *Agrostis, Alopecurus, Arrhenatherum, Bromus, Deschampsia, Festuca, Nardus, Phleum,* and *Poa.* The species strongly depends on ants, especially *Lasius flavus, L. niger,* and *Tetramorium caespitum.*

Note. There may be more than one species involved, or the species shall be subdivided into subspecies, according to Eastop & Hille Ris Lambers (1976). If so the taxon occurring in Scandinavia shall be called *G. eragrostidis* Passerini or *G. utricularia* subsp. *eragrostidis* Passerini.

49. ***Geoica setulosa*** (Passerini, 1860)
Fig. 230.

Tychea setulosa Passerini, 1860: 40.
Geoica herculana Mordvilko, 1935: 215.
Survey: 204.

Apterous viviparous female. Dirty whitish or greenish grey. Antenna about 0.2–0.3 × body; segm. III a little longer than the ultimate segm. in 5-segmented antennae, much longer in 4-segmented antennae; primary rhinaria transverse oval, narrow. Apical segm. of rostrum about 1.8 × 2sht. Anal plate with two longitudinal rows of rather long hairs (Fig. 230). Otherwise as in *utricularia.* 1.2–2.6 mm.

Alate viviparous female: Unknown.

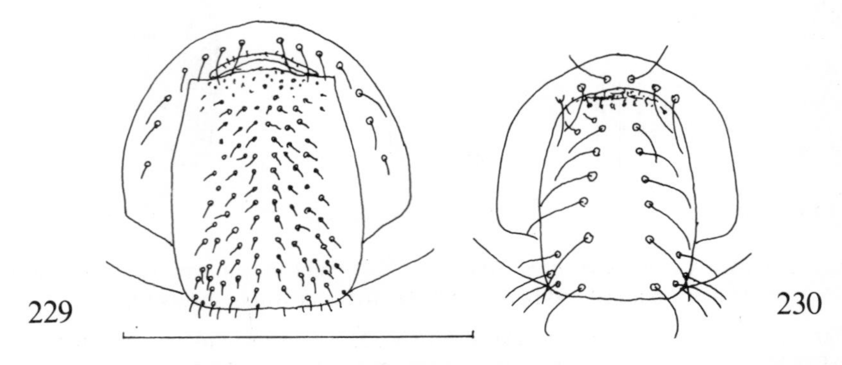

Figs. 229, 230. Anal region of apterae of *Geoica* spp. – 229: *utricularia* (Pass.); 230: *setulosa* (Pass.). The horse-shoe-shaped sclerite in front of anus is abd. tergite VIII; the subrectangular sclerite behind anus is the anal plate (compare with Fig. 223). (Scale 0.5 mm).

Distribution. In Denmark found on *Corynephorus canescens* at Vig Lyng, Odsherred (NWZ); in Sweden known from Sk., Sm., and Öl. (Danielsson), whereas records from Ög. and Stockholm given by Ossiannilsson (1959) refer to *G. squamosa* Hart sensu Tullgren = *G. utricularia;* not in Norway or Finland. – Great Britain, Germany (Naumburg and S Germany), Poland, Italy, and Turkey; not recorded from Russia. Mordvilko described *G. herculana* on the basis of material from Italy and found additional material in Warszawa, Poland.

Biology. The species is anholocyclic, at least in N & C Europe, and alatae are here absent or extremely rare. It feeds on roots of grasses, e.g. *Agrostis, Alopecurus, Corynephorus, Festuca,* and *Holcus.* It is visited and taken care of by ants, especially *Lasius flavus,* which bring the aphids to the nests for overwintering.

SUBTRIBE FORDINA

Primary rhinaria not surrounded by short hairs. Apterae without wax gland plates.
With three genera in Scandinavia (see key on p. 191).

Genus *Paracletus* von Heyden, 1837

Paracletus von Heyden, 1837: 295.
Type-species: *Paracletus cimiciformis* von Heyden, 1837.
Survey: 333.

Antennae of all morphs 6-segmented; primary rhinaria not surrounded by rings of short hairs; secondary rhinaria absent from apterae; alatae with numerous very small, transverse oval or subcircular secondary rhinaria dispersed at random on the undersides of ant. segm. III–VI (Fig. 232). Wings roof-like in repose; media of fore wing unbranched; cubital veins leaving the main vein from the same point. Hind wing with two widely separated oblique veins. Wax gland plates absent from all morphs.
With four species in the world, one species in Scandinavia.

50. ***Paracletus cimiciformis*** von Heyden, 1837
Figs. 231–232.

Paracletus cimiciformis von Heyden, 1837: 294.
Pemphigus pallidus Derbès, 1869: 105.
Survey: 334.

Apterous viviparous female. Shining white or yellowish white. Body flattened, especially the margins, with longitudinal dorsal rows of sunken points (muscle sclerites), and a ventral groove between middle coxae. Cuticle reticulate, with numerous short hairs. Genital aperture of crucial shape because of presence of langitudinal incisures at the posterior edge of the genital plate and in front of the anal plate. Eyes usually with more than three ommatidia. Antenna about 0.3 × body; segm. III about 2–3 × II and

1.1–1.5 × IV; V a little shorter than IV and about as long as VI. Rostrum reaching past hind coxae. Legs strongly built, hind tibiae curved and very long. 2.4–3.5 mm.

Alate viviparous female. Head and thorax dark brown. Abdomen yellowish brown, with somewhat darker central part and anal area. Antenna (Fig. 232) 0.45 × body; processus terminalis extremely short; secondary rhinaria as described in generic diagnosis, on III: 39–50, IV: 22–40, V: 7–17, VI: 1–5. About 2.6 mm.

Distribution. In Sweden found in Sk., Hall., Öl., Gtl. and Ög.; in Finland in Ab and N; not in Denmark or Norway. – Widespread in Europe, incl. the British Isles, France, the Netherlands, Germany (incl. N Germany), Poland, and NW & W Russia; south to the Mediterranean (Spain, Italy, Cyprus). N Africa. Asia: Middle East, east to Iran; W Siberia; C Asia; also recorded from Korea.

Biology. The species is holocyclic in warm climates, with host-alternation between *Pistacia* (I) and roots of grasses (II). It produces flat folds of the edges of leaves of *Pistacia terebinthus* in the Mediterranean. It is anholocyclic in Europe north of the Mediterranean region and in C Asia. It has been recorded from roots of *Agropyrum, Triticum, Hordeum, Festuca, Agrostis, Poa,* and other grasses and also from dicotyledones (*Taraxacum, Lotus* a.o.) in ants' nests. Adlerz (1913) and Tullgren (1925) found in Sweden that the aphids occur in nests of *Tetramorium caespitum.* They usually do not sit on roots, but are found, often in large numbers, in the soil in subterranean galleries built by the ants. If the nest is disturbed the aphids escape into unspoiled galleries. They move like bugs. The ants apparently keep the aphids as cattle and take deep care of them. Adlerz suggested that the ants feed them in the same way as they feed each other. The aphids produce only small amounts of honeydew. The ants benefit from the aphids mostly by licking them according to observations made by Adlerz (1913) and confirmed by Zwölfer (1958). These authors also stated that *Paracletus* contrary to all other aphids is able to absorb fluids directly, and not only from plant tissues.

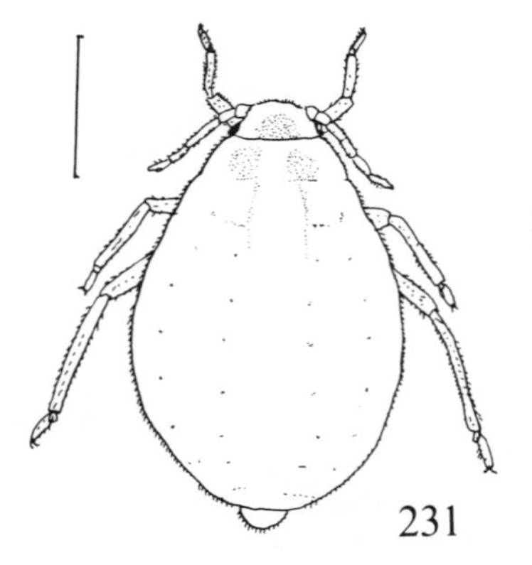

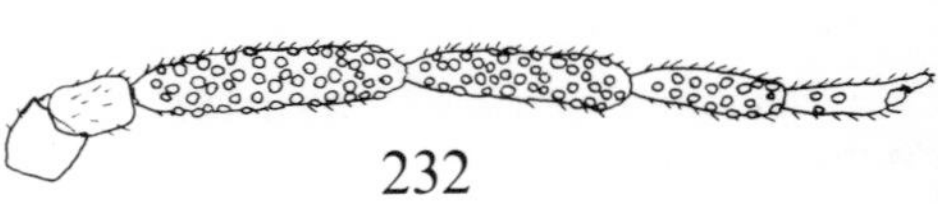

Figs. 231. 232. *Paracletus cimiciformis* v. Heyd. – 231: apterous viviparous female. (Scale 1 mm); 232: antenna of alate viviparous female. (After Mordvilko, redrawn).

Note. Tullgren (1925) gave thorough descriptions and a German translation of the biological observations made by Adlerz (1913).

Genus *Forda* von Heyden, 1837

Forda von Heyden, 1837: 291.
Type-species: *Forda formicaria* von Heyden, 1837.
Survey: 200.

Virginogeniae with 5-segmented antenna. Wax gland plates absent; wax powdering absent or present only on cauda. Ant. segm. II shorter than III. Primary rhinaria without thick, sclerotized rims, not surrounded by short hairs. Media of fore wing unbranched; bases of cubital veins slightly separated. Oblique veins of hind wing separated at bases. Two rudimentary gonapophyses.

With 8 species in the world, 2 species in Scandinavia. Populations in S Europe holocyclic and host-alternating between *Pistacia* (I) and roots of grasses (II). Populations in Scandinavia and other regions, where *Pistacia* does not grow, exclusively anholocyclic on grasses.

Key to species of *Forda*

Apterous viviparous females

1 Head with small spinules not forming a reticulate pattern (Fig. 233). Primary rhinarium on ant. segm. V transverse oval (Fig. 235). Apical segm. of rostrum about 1.5 × 2sht. Colour usually greenish. .... 51. *formicaria* v. Heyden
– Head with small spinules usually forming a more or less distinct reticulate pattern (Fig. 234). Primary rhinarium on ant. segm. V roundish (Fig. 236). Apical segm. of rostrum about 1.3 × 2sht. Colour usually ochreous or yellowish. .... 52. *marginata* Koch

Alate viviparous females

1 Ant. segm. III with 25 or more secondary rhinaria. .... 51. *formicaria* v. Heyden
– Ant. segm. III with less than 25 secondary rhinaria. .... 52. *marginata* Koch

51. ***Forda formicaria*** von Heyden, 1837
Plate 1: 10. Figs. 233, 235.

Forda formicaria von Heyden, 1837: 292. – Survey: 200.

Apterous viviparous female. Pale whitish green, dark green, or bluish green. Head with numerous spinules forming rows or an irregular, not reticulate pattern (Fig. 233). Body hairs long and pointed in some individuals, short, blunt, and sometimes spatulate in others. Eyes with three ommatidia. Antenna about 0.3–0.4 × body; segm. III about 3 × II or longer, about twice as long as IV or longer, usually longer than IV+V; processus terminalis about 0.2 × Va; primary rhinarium on V larger than that on IV, broadly transverse oval, covering more than half of the circumference of the segment (Fig. 235). Rostrum reaching past hind coxae; apical segm. 1.5–1.6 × 2sht. Anus in terminal position, surrounded by hairs. 1.9–3.3 mm.

Alate viviparous female. Abdomen greenish with thin, dark, dorsal cross bands and marginal sclerites; the cross bands may be obliterated in the middle. Antenna about 0.4 × body; segm. III 1.2 × IV+V or a little longer; IV about as long as V; III with 25–42 circular or oval secondary rhinaria, IV with 1–5, V with 0(–1). 2.3–2.8 mm.

Distribution. In Denmark widespread and common (EJ, WJ, NWJ, NEZ); in Sweden common in the south, north to Lu.Lpm.; in Norway known from AAi and HOy; in Finland from Ab and N. – Common all over Europe, incl. the British Isles, N Germany, Poland, and Russia, south to the Mediterranean Sea; in Asia from the Middle East to Iran and C Asia, in Siberia east to the Pacific Ocean, and in Manchuria; N Africa; widespread in the USA and Canada.

Biology. The species is holocyclic in S Europe. The primary host is *Pistacia terebinthus,* whereupon large, crescent-shaped galls are formed by folding of the edges of the leaflets. In N & C Europe exclusively anholocyclic on roots of the secondary hosts, e.g. *Agropyrum, Agrostis, Bromus, Dactylis, Deschampsia, Elymus, Festuca, Hordeum, Nardus, Poa, Secale,* and *Triticum.* It has been recorded also from *Carex* and *Luzula,* but the real hosts are grasses. The aphids are attended by ants and hibernate in ants' nests, especially nests of *Lasius flavus.*

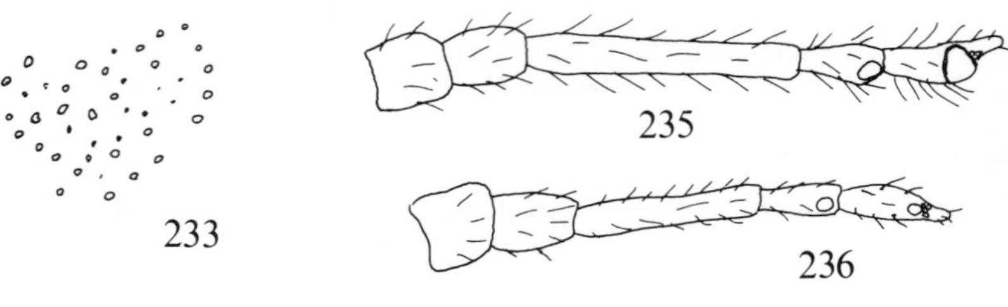

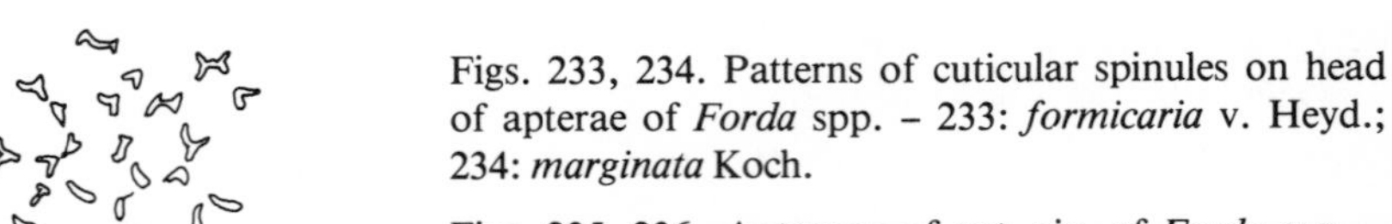

Figs. 233, 234. Patterns of cuticular spinules on head of apterae of *Forda* spp. – 233: *formicaria* v. Heyd.; 234: *marginata* Koch.

Figs. 235, 236. Antennae of apt. viv. of *Forda* spp. – 235: *formicaria* v. Heyd.; 236: *marginata* Koch.

52. ***Forda marginata*** Koch, 1857
Figs. 234, 236–239.

Forda marginata Koch, 1857: 311. – Survey: 200.

Apterous viviparous female. Ochreous, yellow, or greenish yellow. Head and prothorax with spinules forming a more or less distinct reticulate pattern (Fig. 234). Antenna about 0.25 × body; segm. III as long as or a little shorter than IV+V; processus terminalis 0.2–0.4 × Va; primary rhinarium on V roundish, usually of about the same size as that on IV (Fig. 236). Apical segm. of rostrum about 1.2–1.3 × 2sht. Otherwise rather similar to *formicaria*. 1.6–2.9 mm.

Alate viviparous female. Abdomen green with brown, dorsal sclerites or cross bands. Antenna (Fig. 239) about 0.3 × body; secondary rhinaria rather large, on segm. III: 10–21, IV: 1–3, V: 0–1. Otherwise rather similar to *formicaria*. 1.9–2.6 mm.

Distribution. In Denmark collected in a trap at Tåstrup near Copenhagen (NEZ) and from ants' nest on Samsø (EJ); in Sweden recorded from Sk., Sm., Öl., and Upl.; known from Norway; in Finland at Åminsby (N). – Widespread in Europe, incl. the British Isles, N Germany, Poland, and NW Russia, south to the Mediterranean Sea, Crimea, and the Caucasus region; Middle East, Siberia, and N Africa; N America.

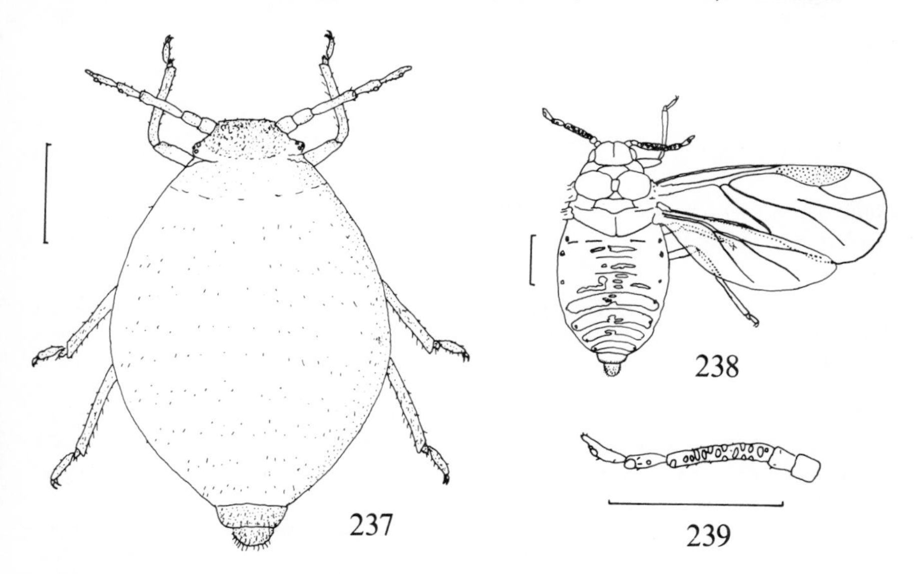

Fig. 237. *Forda marginata* Koch, apt. viv.

Figs. 238, 239. Same, al. viv. (virginopara). – 238: body and wings; 239: antenna. (After Zwölfer). (All scales 0.5 mm).

Biology. The species is holocyclic in S Europe. The primary hosts are *Pistacia terebinthus* and *P. mutica*. The galls are sack-shaped foldings of the edges of the leaflets. The species is exclusively anholocyclic on grass roots in N & C Europe and overwinters in nests of *Lasius* spp. It is attended also by other ant genera. Among the hosts mentioned in literature are *Agrostis, Bromus, Calamagrostis, Corynephorus, Festuca, Hordeum, Poa, Secale,* and *Triticum;* in Finland collected on *Alopecurus*. The species seems to prefer dry and warm soil according to Zwölfer (1958), whereas *F. formicaria* can be found also in wet areas.

Genus *Smynthurodes* Westwood, 1849

Smynthurodes Westwood, 1849: 420.
Type-species: *Smynthurodes betae* Westwood, 1849.
Tullgrenia van der Goot, 1912: 96.
Type-species: *Tychea phaseoli* Passerini, 1860
= *Smynthurodes betae* Westwood, 1849.
Survey: 407.

Virginogeniae with 5-segmented antennae; segm. II relatively long. Primary rhinaria with thick, sclerotized rims, not surrounded by short hairs. Two rudimentary gonapophyses.

One species in the world.

53. ***Smynthurodes betae*** Westwood, 1849
Figs. 240–242.

Smynthurodes betae Westwood, 1849: 420.
Tychea phaseoli Passerini, 1860: 39.
Survey: 407.

Apterous viviparous female. Dirty yellowish white; head, prothorax, antennae, and legs brownish. Slightly wax powdered. Body almost globular, with numerous fine hairs placed in 5–7 transverse rows on each of the abd. segm. Eyes with three or more ommatidia. Antenna 5-segmented, 0.25–0.3 × body; segm. II longer than IV and sometimes even longer than III (Fig. 241); V about as long as III or longer; primary rhinarium on IV circular or subcircular (Fig. 242), that on V transverse oval; III sometimes with a few secondary rhinaria. Rostrum reaching hind coxae; apical segm. about 1.0 × 2sht. Anus in dorsal position. 1.5–2.5 mm.

Alate viviparous female. Abdomen with dark dorsal cross bars. Antenna 5- or 6-segmented, 0.3–0.4 × body; segm. III much longer than II; secondary rhinaria rather large, transverse oval or subcircular, sometimes of irregular shape; in 6-segmented antennae 6–10 secondary rhinaria on III, 2–4 on IV, and 0–1 on V. Media of fore wing unbranched; cubital veins usually slightly separated at bases. Bases of oblique veins of hind wing close to each other, and basal part of the distal one obliterated. 1.9–2.3 mm.

Distribution. Rather few records from Scandinavia, but possibly overlooked; apparently only south of latitude 58° N; in Denmark a few records from potato on Funen (F); in Sweden found by Tullgren (1925) on *Phaseolus* at Landskrona (Sk.), recently also recorded by Danielsson from Sm. and Öl.; not in Norway or Finland. – Almost all over the world; widely distributed in Europe, from S Scandinavia south to the Mediterranean.

Biology. The species is polyphagous. It has been found on roots of several herbaceous dicotyledones, e.g. *Artemisia, Arctium,* and other Compositae, *Phaseolus, Vicia, Trifolium,* and other Leguminosae, *Solanum tuberosum* and *S. nigrum, Beta, Atriplex, Cerastium, Brassica, Capsella,* and *Plantago.* Occurrence on grass-roots *(Festuca rubra)* is reported by Zwölfer (1958), but may be accidental as the species contrary to most other Fordinae seems to infest dicotyledones only. The species is holocyclic in S Europe, with *Pistacia mutica* as the primary host as stated by Mordvilko (1935). The galls on *Pistacia* are formed by folding of the edges of the leaflets. In N & C Europe it is anholocyclic, and here alatae seem to be rare. Mordvilko (1935) described individuals which are alatiform, with rather large compound eyes, 6-segmented antennae, and sometimes rudimentary wings, and stated that they occurred rather frequently. The colonies are attended by ants. They often occur in nests of *Lasius flavus* during the winter according to observations made by Zwölfer in S Germany. The aphids are brought into the nests by the ants in November, where they are found on roots in the beginning

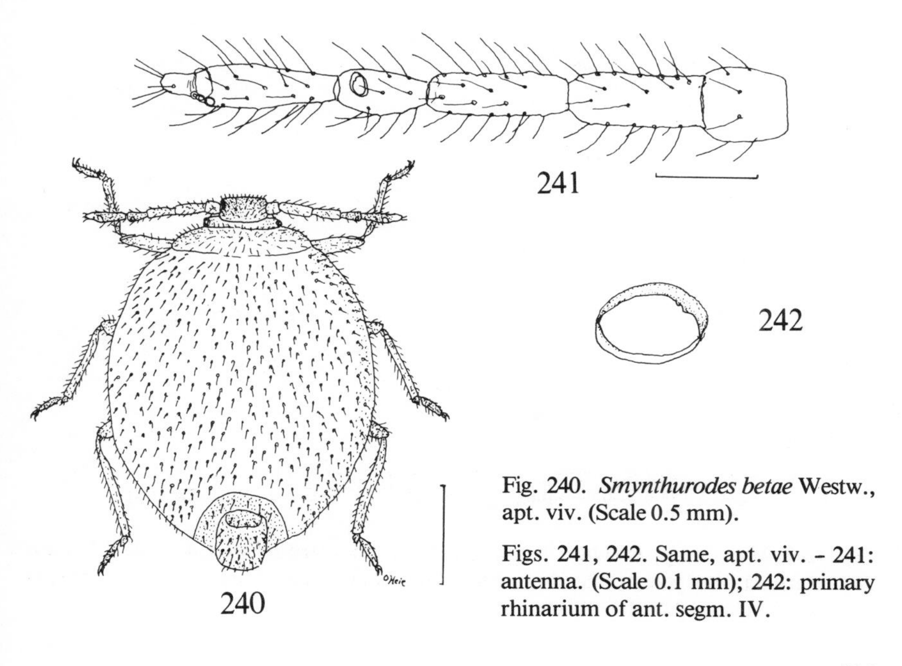

Fig. 240. *Smynthurodes betae* Westw., apt. viv. (Scale 0.5 mm).

Figs. 241, 242. Same, apt. viv. – 241: antenna. (Scale 0.1 mm); 242: primary rhinarium of ant. segm. IV.

of the winter, later in special earth chambers. From April to November most colonies are found on roots outside the nests.

This aphid species is not an important pest in Scandinavia in spite of its occurrence on cultivated plants as potatoes, beets, and beans.

TRIBE MELAPHIDINI

Apterae with one-segmented tarsi. Secondary rhinaria of alatae ringlike. Media of fore wing with one fork. Primary host of holocyclic species *Rhus*.

Genus *Melaphis* Walsh, 1867

Melaphis Walsh, 1867: 281.
Type-species: *Byrsocrypta rhois* Fitch, 1866.
Survey: 276.
One species in the world.

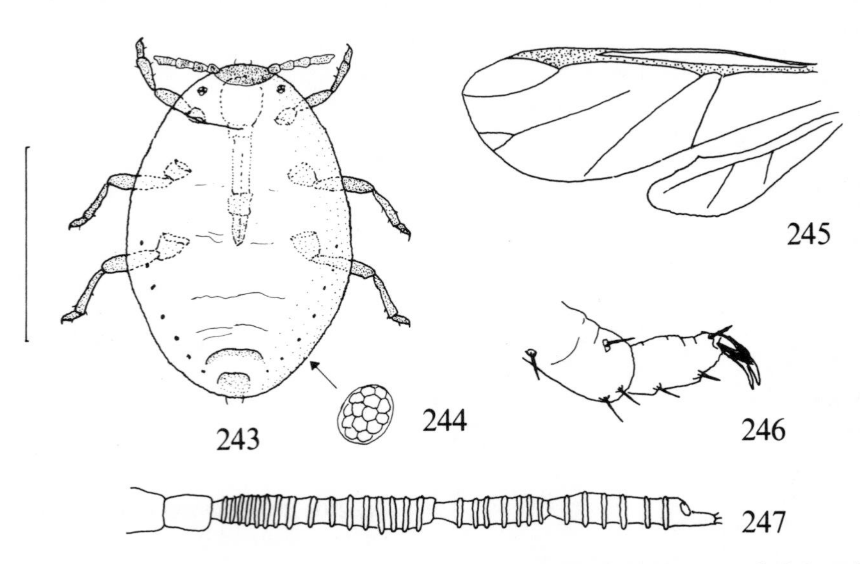

Figs. 243–247. *Melaphis rhois* (Fitch). – 243: apt. viv. (from Sweden, scale 0.5 mm); 244: marginal wax gland plate; 245: wings; 246: hind tarsus of apt. viv.; 247: antenna of al. viv. (245 and 247 after Baker, redrawn).

54. ***Melaphis rhois*** (Fitch, 1866)
Figs. 243–247.

Byrsocrypta rhois Fitch, 1866: 73. – Survey: 276.

Apterous viviparous female on secondary host. Body globular or oval. Pale. Antennae and legs brownish. Sometimes with wax powder produced by facetted, indistinctly visible, dorsal and marginal gland plates on the abdomen (Fig. 244). With rather few short hairs on body and antennae. Antenna 4- or 5-segmented, 0.16–0.18 × body; processus terminalis very short, about 0.3 × base of distal segment; segm. II in 5-segmented antennae longer than III and about as long as IV; distal segm. longer than the other segments. Secondary rhinaria absent. Rostrum reaching middle coxae (in large specimens) or hind coxae (in small specimens). Legs short, with one-segmented tarsi (Fig. 246). 0.8–1.2 mm.

Alate viviparous female on primary host. Antenna 5- or 6-segmented, 0.4 × body; secondary rhinaria narrow, ringlike (Fig. 247); in 6-segmented antennae: III: 10, IV: 5, V: 6, VI: 7, in 5-segmented antennae: III: 16–17, IV: 7–8, V: 6–9; segm. III shorter than IV+V both in 5- and 6-segmented antennae; processus terminalis 0.09–0.11 × base of distal segm. Apical segment of rostrum 0.6–0.7 × 2sht. Media in fore wing with one fork; cubital branches separated; pterostigma rather long drawn out, with concave posterior margin (Fig. 245). Hind wing with two oblique veins. About 1.6 mm.

Distribution. In Sweden found in Sk. and Sm. (on mosses); not in Denmark, Norway, or Finland. – The species is of nearctic origin, widespread in the USA and Canada, also known from Europe, e.g. England; not recorded from N Germany, Poland, and Russia.

Biology. In N America the species is dioecious, with sumach *(Rhus glabra, R. typhina)* as the primary host and mosses as the secondary hosts. Small, sac-like galls are produced on the upper surface of leaves of the primary host. In Europe it lives on mosses, only, and seems to be rare.

				DENMARK											
		Germany	G. Britain	SJ	EJ	WJ	NWJ	NEJ	F	LFM	SZ	NWZ	NEZ	B	Sk.	Bl.
Mindarus abietinus Koch	1	●	●		●		●	●							●	
M. obliquus (Chol.)	2	●	●	●	●	●	●	●	●			●	●		●	
Cerataphis orchidearum (Westw.)	3		●												●	
Hamamelistes betulinus (Horv.)	4		●		●			●			●		●		●	●
Hormaphis betulae (Mordv.)	5															
Thelaxes dryophila (Schr.)	6	●	●	●	●	●	●	●	●	●		●	●		●	●
Glyphina betulae (L.)	7	●	●		●	●	●	●	●						●	●
G. schrankiana Börn.	8		●													
Anoecia corni (F.)	9	●	●	●	●		●			●	●		●		●	●
A. major Börn.	10		●												●	
A. nemoralis Börn.	11		●												●	●
A. vagans (Koch)	12	●	●										●		●	
A. zirnitsi Mordv.	13		●		●										●	
A. (Paranoecia) pskovica Mordv.	14		●		●										●	
Eriosoma lanigerum (Hausm.)	15	●	●	●	●	●		●	●	●			●		●	●
E. (Schizoneura) anncharlotteae Dan.	16														●	
E. (S.) lanuginosum (Htg.)	17	●	●		●			●					●			
E. (S.) patchiae (Börn. & Bl.)	18		●		●		●	●					●		●	
E. (S.) sorbiradicis Dan.	19														●	
E. (S.) ulmi (L.)	20	●	●	●	●	●	●	●	●	●	●	●	●		●	●
Colopha compressa (Koch)	21	●	●				●								●	
Kaltenbachiella pallida (Hal.)	22	●	●		●								●		●	
Tetraneura longisetosa (Dahl)	23															
T. ulmi (L.)	24	●	●	●	●	●	●	●	●	●	●	●	●	●	●	
Pachypappa populi (L.)	25															
P. tremulae (L.)	26	●	●	●			●	●							●	
P. vesicalis Koch	27	●													●	
P. warschavensis (Nas.)	28														●	
Pachypappella lactea (Tullgr.)	29														●	●
Gootiella tremulae Tullgr.	30												●		●	
Prociphilus bumeliae (Schr.)	31		●				●	●							●	●
P. fraxini (F.)	32	●	●				●						●		●	
P. (Stagona) pini (Burm.)	33		●				●	●					●		●	
P. (S.) xylostei (DeGeer)	34	●	●			●	●		●	●			●	●	●	
Mimeuria ulmiphila (d. Gu.)			●													
Thecabius affinis (Kalt.)	35	●	●		●	●	●	●	●			●	●	●	●	●
T. (P.) lysimachiae Börn.	36		●		●								●		●	
T. (P.) auriculae (Murr.)			●													

SWEDEN

	Hall.	Sm.	Öl.	Gtl.	G. Sand.	Ög.	Vg.	Boh.	Dlsl.	Nrk.	Sdm.	Upl.	Vstm.	Vrm.	Dlr.	Gstr.	Hls.	Med.	Hrj.	Jmt.	Ång.	Vb.	Nb.	Ås. Lpm.	Ly. Lpm.	P. Lpm.	Lu. Lpm.	T. Lpm.
1	●		●	●		●	●					●	●		●													
2																												
3																												
4		●					●				●	●				●										●		
5																												
6	●	●	●	●		●	●	●				●																
7	●	●	●	●			●				●	●	●	●	●							●				●		
8												●																
9	●	●	●				●			●		●										●						
10																												
11		●	●																									
12	●	●	●			●	●					●																
13		●	●																									
14		●																										
15			●			●	●		●																			
16																												
17																												
18							●																					
19												●																
20	●	●	●	●		●	●	●	●		●	●	●	●	●		●	●										
21		●	●									●																
22			●	●		●	●																					
23												●																
24	●	●	●			●	●	●	●		●	●		●	●		●											●
25	●	●				●	●			●	●	●	●	●	●													
26	●	●	●	●		●	●				●	●	●	●	●				●	●	●		●	●				●
27		●	●			●					●	●																
28																												
29		●	●			●	●		●			●	●	●	●					●		●	●					
30		●	●			●																						
31	●		●				●				●	●																
32												●																
33		●	●	●			●					●		●														
34	●	●	●	●		●	●	●			●	●		●	●		●	●			●	●						
35	●	●	●	●		●	●	●			●	●		●				●					●			●		
36												●																

		NORWAY														
		Ø + AK	HE (s + n)	O (s + n)	B (ø + v)	VE	TE (y + i)	AA (y + i)	VA (y + i)	R (y + i)	HO (y + i)	SF (y + i)	MR (y + i)	ST (y + i)	NT (y + i)	Ns (y + i)
Mindarus abietinus Koch	1	◗														
M. obliquus (Chol.)	2					N	O	R	W	A	Y					
Cerataphis orchidearum (Westw.)	3															
Hamamelistes betulinus (Horv.)	4															
Hormaphis betulae (Mordv.)	5															
Thelaxes dryophila (Schr.)	6	◗						◖		◖	◖					
Glyphina betulae (L.)	7					N	O	R	W	A	Y					
G. schrankiana Börn.	8										◗					
Anoecia corni (F.)	9					N	O	R	W	A	Y					
A. major Börn.	10															
A. nemoralis Börn.	11															
A. vagans (Koch)	12															
A. zirnitsi Mordv.	13															
A. (Paranoecia) pskovica Mordv.	14															
Eriosoma lanigerum (Hausm.)	15									◖			◖			
E. (Schizoneura) anncharlotteae Dan.	16															
E. (S.) lanuginosum (Htg.)	17															
E. (S.) patchiae (Börn. & Bl.)	18	◗														
E. (S.) sorbiradicis Dan.	19															
E. (S.) ulmi (L.)	20	●	◖	◖	◖	●	◖		◖	◖	◖	◗	◖		◗	◖
Colopha compressa (Koch)	21															
Kaltenbachiella pallida (Hal.)	22															
Tetraneura longisetosa (Dahl)	23															
T. ulmi (L.)	24	●			◖		●			◖	●			◗		
Pachypappa populi (L.)	25					N	O	R	W	A	Y					
P. tremulae (L.)	26	◗	●	●	◖											
P. vesicalis Koch	27															
P. warschavensis (Nas.)	28															
Pachypappella lactea (Tullgr.)	29															
Gootiella tremulae Tullgr.	30					N	O	R	W	A	Y					
Prociphilus bumeliae (Schr.)	31															
P. fraxini (F.)	32					N	O	R	W	A	Y					
P. (Stagona) pini (Burm.)	33	◗														
P. (S.) xylostei (DeGeer)	34														◗	
Mimeuria ulmiphila (d. Gu.)																
Thecabius affinis (Kalt.)	35	●	◖	◖	●	●			◖		◖		◖		◖	
T. (P.) lysimachiae Börn.	36															
T. (P.) auriculae (Murr.)																

					FINLAND																				USSR		
	Nn (ø + v)	TR (y + i)	F (v + i)	F (n + ø)	Al	Ab	N	Ka	St	Ta	Sa	Oa	Tb	Sb	Kb	Om	Ok	Ob S	Ob N	Ks	LkW	LkE	Le	Li	Vib	Kr	Lr
1						●	●			●	●	●						●									
2														●													
3							●																				
4							●					●							●								
5							●																				
6					●	●	●			●																	
7					●		●			●	●			●			●	●							●		
8							●			●	●				●												
9						●	●	●		●	●														●		
10																											
11																											
12																											
13																											
14																											
15																											
16																											
17																											
18																											
19																											
20					●	●	●	●	●	●	●	●		●		●										●	
21						●	●			●				●													
22										●																	
23																											
24		◗			●	●	●	●	●	●	●																
25						●	●	●		●					●										●		
26					●	●	●				●			●										●			
27																											
28																											
29					●	●	●			●			●	●										●			
30							●																				
31																											
32																											
33							●				●			●													
34					●	●	●	●		●	●						●										
35						●	●			●		●															
36																											

				DENMARK												
		N. Germany	G. Britain	SJ	EJ	WJ	NWJ	NEJ	F	LFM	SZ	NWZ	NEZ	B	Sk.	Bl.
Pemphigus borealis Tullgr.	37		●													
P. bursarius (L.)	38	●	●	●	●	●	●	●		●		●	●	●	●	●
P. fuscicornis (Koch)	39				●		●		●				●		●	
P. gairi Stroyan			●													
P. phenax Börn. & Bl.	40		●									●				●
P. populi Courch.	41		●													
P. populinigrae (Schr.)	42	●	●		●		●	●	●	●	●		●	●	●	
P. protospirae Licht.	43	●	●												●	
P. saliciradicis (Börn.)	44		●													
P. spyrothecae Pass.	45	●	●		●	●	●	●	●	●	●	●	●	●	●	
P. trehernei Fost.			●													
Aploneura lentisci (Pass.)	46		●				●									
Baizongia pistaciae (L.)	47		●												●	
Geoica utricularia (Pass.)	48	●	●		●	●	●	●				●			●	
G. setulosa (Pass.)	49		●									●			●	
Paracletus cimiciformis v. Heyd.	50	●	●												●	
Forda formicaria v. Heyd.	51	●	●		●	●	●						●		●	●
F. marginata Koch	52	●	●		●								●		●	
Smynthurodes betae Westw.	53	●	●						●						●	
Melaphis rhois (Fitch)	54		●												●	

	Hall.	Sm.	Öl.	Gtl.	G. Sand.	Ög.	Vg.	Boh.	Dlsl.	Nrk.	Sdm.	Upl.	Vstm.	Vrm.	Dlr.	Gstr.	Hls.	Med.	Hrj.	Jmt.	Ång.	Vb.	Nb.	Ås. Lpm.	Ly. Lpm.	P. Lpm.	Lu. Lpm.	T. Lpm.
37						●					●	●				●												
38		●	●			●	●	●			●	●																
39																												
40																												
41		●				●	●																					
42		●	●			●	●	●			●	●																
43			●																									
44																												●
45	●	●	●	●		●	●					●																
46																												
47		●	●								●																	
48		●	●			●						●																
49		●	●									●																
50	●		●	●		●																						
51		●	●					●				●															●	
52		●	●									●																
53		●	●																									
54		●																										

		NORWAY														
		Ø + AK	HE (s + n)	O (s + n)	B (ø + v)	VE	TE (y + i)	AA (y + i)	VA (y + i)	R (y + i)	HO (y + i)	SF (y + i)	MR (y + i)	ST (y + i)	NT (y + i)	Ns (y + i)
Pemphigus borealis Tullgr.	37															
P. bursarius (L.)	38	●			●											
P. fuscicornis (Koch)	39															
P. gairi Stroyan																
P. phenax Börn. & Bl.	40															
P. populi Courch.	41															
P. populinigrae (Schr.)	42								◖							
P. protospirae Licht.	43															
P. saliciradicis (Börn.)	44															
P. spyrothecae Pass.	45	●			◖		●									
P. trehernei Fost.																
Aploneura lentisci (Pass.)	46															
Baizongia pistaciae (L.)	47															
Geoica utricularia (Pass.)	48															
G. setulosa (Pass.)	49															
Paracletus cimiciformis v. Heyd.	50															
Forda formicaria v. Heyd.	51							◗			◖					
F. marginata Koch	52					N	O	R	W	A	Y					
Smynthurodes betae Westw.	53															
Melaphis rhois (Fitch)	54															

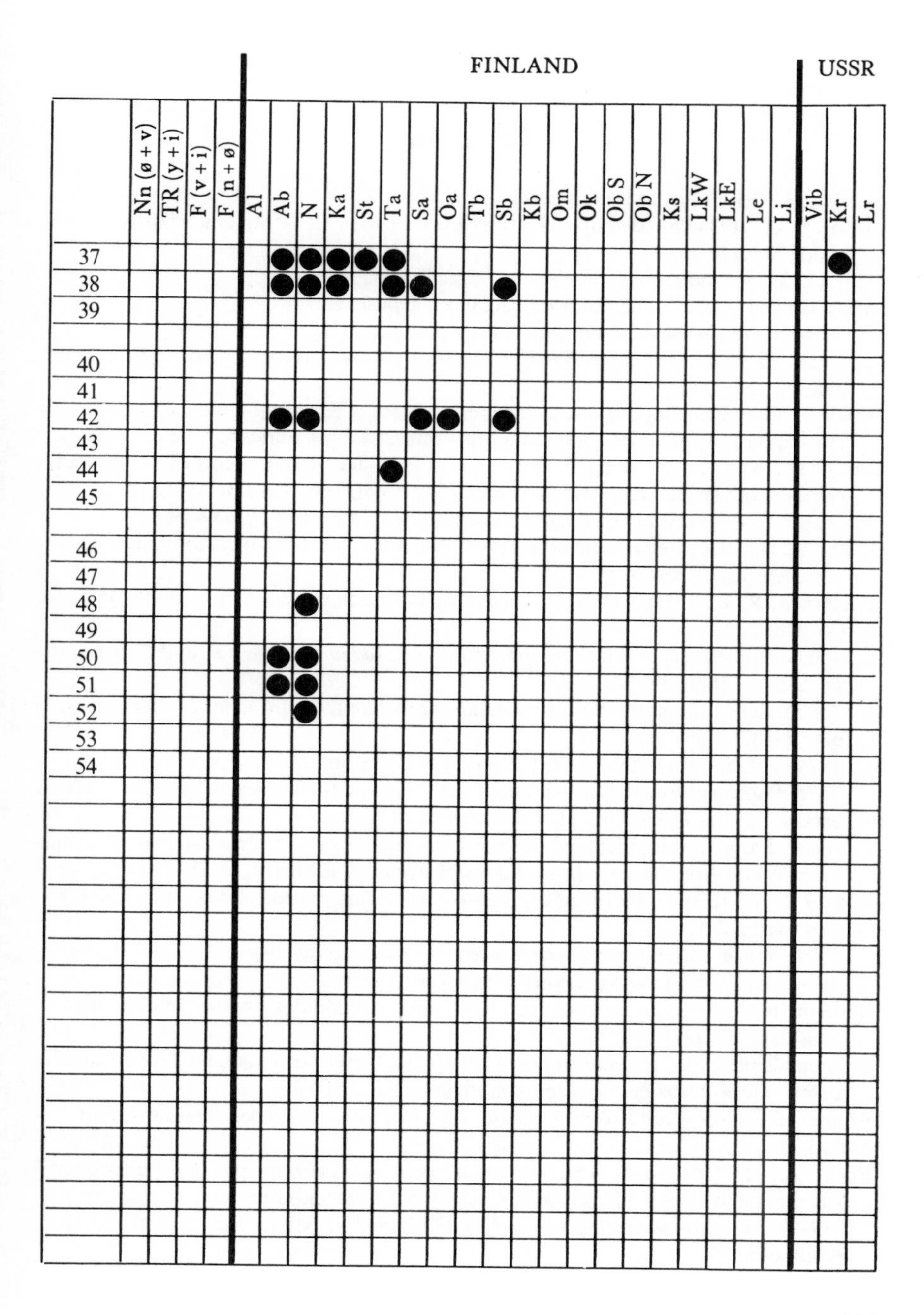

					FINLAND																				USSR		
	Nn (ø + v)	TR (y + i)	F (v + i)	F (n + ø)	Al	Ab	N	Ka	St	Ta	Sa	Oa	Tb	Sb	Kb	Om	Ok	Ob S	Ob N	Ks	LkW	LkE	Le	Li	Vib	Kr	Lr
37						●	●	●	●	●																●	
38						●	●	●		●	●			●													
39																											
40																											
41																											
42						●	●				●	●		●													
43																											
44										●																	
45																											
46																											
47																											
48							●																				
49																											
50						●	●																				
51						●	●																				
52							●																				
53																											
54																											

Literature

Adlerz, G., 1913: Myrornas liv. – Stockholm.

Ahlberg, O., 1934: Skadedjur i Sverige åren 1928–1932. – Statens växtskyddsanst. medd., Stockholm, 7.

Alm, G., 1915: Bladlushärjning på bok. – Ent. Tidskr., 36.

Andersson, H., 1955: Halländska zoocecidier. – Opusc. ent., 20.

– , 1958: Några svenska och nordnorska cecidiefynd. – Ibid., 23.

– , 1959: Några interessantare fynd av svenska växtlöss (Hem. Aphid. et Coccid.). – Ibid., 24: 153–155.

Aoki, S., 1975: Descriptions of the Japanese species of *Pemphigus* and its allied genera (Homoptera: Aphidoidea). – Insecta Matsumurana, New Series 5: 1–63.

Archibald, K. D., 1957: Report of the forest Aphidae (Homoptera) of Nova Scotia together with species and host plant lists. – Ann. ent. Soc. Am., 50: 38–42.

Auclair, J. L., 1963: Aphid feeding and nutrition. – A. Rev. Ent., 8: 439–490.

Baker, A. C., 1920: Generic classification of the hemipterous family Aphididae. – U. S. Dept. Agr. Bull., 826, 93 pp.

Blackman, R. L., 1965: Studies on specificity in Coccinellidae. – Ann. appl. Biol., 56: 336–338.

– , 1967a: The effects of different aphid foods on *Adalia bipunctata* L. and *Coccinella 7-punctata* L. – Ibid., 59: 207–219.

– , 1967b: Selection of aphid prey by *Adalia bipunctata* L. and *Coccinella 7-punctata* L. – Ibid., 59: 331–338.

– , 1974: Aphids. – Invertebrate Types, Ginn & Co., London.

– , 1975: Photoperiodic determination of the male and female sexual morphs of *Myzus persicae*. – J. Insect Physiol., 21: 435–453.

– , 1978: Early development of the parthenogenetic egg in three species of aphids (Homoptera: Aphididae). – Int. J. Insect Morphol. & Embryol., 7: 33–44.

Boas, J. E. V., 1890: Om en Rodlus, *Pemphigus Poschingeri,* paa Ædelgran. – Tidsskr. f. Skovvæsen, 2 A.

– , 1904: Ædelgranfjenderne *Chermes piceae* og *Mindarus abietinus*. – Ibid., 17 B: 1–19.

– , 1924: Dansk Forstzoologi. 2nd edition. Copenhagen.

Bodenheimer, F. S. & Swirski, E., 1957: The Aphidoidea of the Middle East. – Jerusalem. 378 pp.

Bonnemaison, L., 1951: Contribution à l'étude des facteurs provoquant l'apparition des formes ailées et sexuées chez les Aphidinae. – Paris. 380 pp.

Bonnet, C., 1745: Traité d'Insectologie, ou Observations sur les Pucerons. Part I. – Paris.

Borg, Å., 1949a: Några växtsjukdomar och skadegörare på lantbruksväxter i Västergötland 1949. – Växtskyddsnotiser 13, No. 6, Stockholm.

– , 1949b: Blodlusinventeringen i Skåne 1945–1948. – Medd. Stat. växtskyddsanst., Stockholm, 53.

– , 1951: Några växtsjukdomar och skadedjur i Västergötland 1950. – Växtskyddsnotiser, 15.
Börner, C., 1930: Beiträge zur einem neuen System der Blattläuse. – Arch. klass. u. phyl. Ent., 1: 115–180.
– , 1950: Neue europäische Blattlausarten. – Selbstverlag, Naumburg, Saale. 19 pp.
– , 1952: Europae centralis Aphides. – Mitt. Thür. Bot. Ges. Beiheft 3, Weimar. 488 pp.
– & Blunck, H., 1916: Beiträge zur Kenntnis der wandernde Blattläuse Deutschlands. – Mitt. Kais. Biol. Anst. H. 16: 25–49.
– & Heinze, K., 1957: Aphidina – Aphidoidea. – In: P. Sorauer: Handbuch der Pflanzenkrankheiten, 2. Teil, 4. Lief. Homoptera II: 1–402.
Bovien, P., 1939: Skadedyr af særlig interesse. – Tidsskr. f. Planteavl, Copenhagen, 44: 37–49.
– & Thomsen, M., 1945 (new edition: 1950). Haveplanternes Skadedyr og deres Bekæmpelse. – Copenhagen.
Bowers, W. S., Nault, L. R., Webb, R. E. & Dutky, S. R., 1972: Aphid alarm pheromone: Isolation, identification, synthesis. – Science 177: 1121–1122.
Broadbent, L., 1949: Factors effecting the activity of alate of the aphids *Myzus persicae* (Sulzer) and *Brevicoryne brassicae* (L.). – Ann. appl. Biol., 36: 40–62.
– , 1951: Aphid excretion. – Proc. R. ent. Soc. Lond. A, 26: 97–103.
Buckton, G. B., 1876–1883: Monograph of the British Aphides. – London, Ray Soc. I (1876): iii + 193 pp.; II (1879): 176 pp.; III (1881): ii + 142 pp.; IV (1883): ix + 128 pp.
Burmeister, H., 1835: Handbuch der Entomologie 2: 85–95.
Carter, C. I. & Eastop, V. F., 1973: *Mindarus obliquus* (Chol.) (Homoptera, Aphidoidea) new to Britain and records of two other aphids recently found feeding on conifers. – Entomologist's mon. Mag., 108 (1972): 202–204.
Cholodkovsky, N., 1896: Über die auf Nadelhölzern vorkommenden Pemphigiden. – Zool. Anz., 19: 257–260.
Cottier, W., 1953: Aphids of New Zealand. – N. Z. Dept. Sci. Industr. Res. Bull., 106: 1–382.
Courchet, L., 1879: Etude sur les galles produites par les Aphidiens. – Mém. Acad. Montpellier.
Dahl, F., 1912: Über die Fauna des Plagefenngebietes: Schnabelkerfe. – Beitr. Naturdenkmalpflege, Berlin, 3: 431–435.
Dahlbom, A. G., 1851: Anteckningar öfver Insekter som blifvit observerade på Gotland och i en del af Calmare län, under sommaren 1850. – K. svenska Vetensk-Akad. Handl., 1850.
Danielsson, R., 1974: New records of Swedish aphids (Hem. Hom. Aphidoidea). – Ent. Tidskr., 95: 64–72.
– , 1976: Gallbildande bladlöss på ask och poppel i Sverige. – Entomologen, Lund, 5: 1–14.
– , 1979: The genus *Eriosoma* Leach in Sweden, with descriptions of two new species. Studies on Eriosomatidae I (Homoptera, Aphidoidea). – Ent. scand., 10: 193–208.

Davatchi, A., 1948: Les pucerons du peuplier. – Entomol. Phytopathol. Appl. Teheran, 6/7: 1–5.
Davies, W. M., 1936: Studies on the aphids infecting the potato crop. V. Laboratory experiments on the effect of wind velocity on the flight of *Myzus persicae* Sulz. – Ann. appl. Biol., 23: 401–408.
DeGeer, C., 1755: Mémoires pour servir à l'Histoire des Insectes. – Stockholm.
– , 1773: Des Pucerons. Mémoires pour servir à l'Histoire des Insectes. Stockholm. 3: 27–77, 565–629.
Derbès, A., 1869: Observations sur les Aphidiens qui font les galles des Pistachiers. – Ann. Soc. Nat. Zool. Ser. V., 11: 93–107.
Dixon, A. F. G., 1973: Biology of aphids. – Studies in Biology No. 44. London. 58 pp.
Dolgova, L. P., 1970: The heteroecious life cycle of *Pemphigus borealis* Tullg. (Homoptera, Aphidoidea). – Ent. Rev., 49: 17–23.
Doncaster, J. P., 1961: Francis Walker's Aphids. – British Museum, London, viii + 165 pp.
– , 1973: G. B. Buckton's works on Aphidoidea (Hemiptera). – Bull. Brit. Mus. (Nat. Hist.) Ent., 28(2): 23–109.
Eastop, V. F., 1958: A study of the Aphididae (Homoptera) of East Africa. – H. M. S. O., London. 126 pp.
– , 1961: A study of the Aphididae (Homoptera) of West Africa. – Brit. Mus., London. 93 pp.
– , 1966: A taxonomic study of Australian Aphidoidea (Homoptera). – Aust. J. Zool., 14: 399–592.
– & Emden, H. F. van, 1972: The insect material. – Chapter 1 of H. F. van Emden (ed.): Aphid Technology. – Academic Press, London and New York, pp. 1–45.
– & Hille Ris Lambers, D., 1976: Survey of the World's aphids. – Junk, the Hague. 573 pp.
– & Tanasijević, N., 1968: Aphid records from Yugoslavia. – Entomologist's mon. Mag., 104: 55–57.
Emden, H. F. van, 1973: Aphid host plant relationships. – In: A. D. Lowe (ed.): Perspectives in aphid biology. – Ent. Soc. New Zeal. Bull., 2: 54–64.
Essig, E. O., 1953a: Some new and noteworthy Aphidae from western and southern South America (Hemiptera – Homoptera). – San Francisco. 164 pp.
– , 1953b: Aphid miscellany. – Pan-Pacif. Ent., 29: 1–13.
– , 1956: Insects of Micronesia. Homoptera: Aphididae. – Insects Micronesia, 6: 15–37.
Fabricius, J. C., 1775: Systema Entomologiae. – Flensburgi et Lipsiae, xxviii + 832 pp.
– , 1777: Genera Insectorum. – Chilonii. Praefatio 1776. Editio 1777. 310 pp.
– , 1781: Species Insectorum. – Hamburgi et Kilonii. Vol. 2. 517 pp.
– , 1787: Mantissa Insectorum. 2: 1–382.
– , 1794: Entomologia Systematica Emendata et Aucta. 4: 1–472. – Hafniae.
– , 1803: Systema Rhyngotorum. X + 314 pp.
Fitch, A., 1859: Fifth report on the noxious, beneficial, and other insects of the State of New York. – N. Y. State Agr. Soc. Transact., 18: 843.

– , 1866: Sumac gall-aphis; *Byrsocrypta rhois* (Order Homoptera, Family Aphidae). – J. N. Y. State Agr. Soc., 17: 73.
Fjelddalen, 1964: Aphids recorded on cultivated plants in Norway 1946–62. – Norsk ent. Tidsskr., 12: 259–295.
Fonscolombe, E. L. J. H. Boyer de, 1841: Descriptions des pucerons qui se trouvent aux environs d'Aix. – Ann'ls Soc. ent. Fr. 10: 157–198.
Foster, W. A., 1975: A new species of *Pemphigus* Hartig (Homoptera: Aphidoidea) from Western Europe. – J. Ent. (B), 44 (3): 255–263.
Furk, C. & Prior, R. N. B., 1975: On the life cycle of *Pemphigus (Pemphiginus) populi* Courchet, with a key to British species of *Pemphigus* Hartig (Homoptera: Aphidoidea). – J. Ent. (B), 44(3): 265–280.
Geer, C. de: See DeGeer, C.
Gertz, O., 1918: Skånes zoocecidier. Ett bidrag till kännedomen om Sveriges gallbildande flora och fauna. – K. Fysiogr. Sällsk. Handl. N. F., 29, No. 26.
Gittins, A. R., Bishop, G. W., Knowlton, G. F. & Parker, E. J., 1976: An annotated list of the aphids of Idaho. – Agr. Exp. Sta. Univ. Idaho, Res. Bull. No. 95. 48 pp.
Gleiss, H. G. W., 1967: Der derzeitige Stand unseres Wissens über die Blattlausfauna von Schleswig-Holstein und Hamburg (Homoptera: Aphidoidea). – Faun.-Ökol. Mitt., 3: 124–163.
Goot, P. van der, 1912: Über einige noch nicht oder nur unvollständig beschriebene Blattlausarten. – Tijdschr. Ent., 55: 58–96.
– , 1913: Zur Systematik der Aphiden. – Ibid., 56: 69–155.
– , 1915: Beiträge zur Kenntnis der holländischen Blattläuse. – Haarlem und Berlin. 600 pp.
– , 1917: Zur Kenntnis der Blattläuse Javas. – Contrib. à la Faune des Indes Néerland., 1 (Fasc. 3): 1–301.
– , 1918: Introduction and notes *in:* B. Das: The Aphididae of Lahore. – Mem. Indian Mus., 6: 135–274.
Guercio, G. del, 1909: Intorno a due nuove generi e a tre specie nuove di afidi di California. – Riv. Patol. Veg., 3: 328–332.
Hagen, K. S. & Bosch, R. van den, 1968: Impact of pathogens, parasites and predators on aphids. – A. Rev. Ent., 13: 325–384.
Hart, C. A., 1894: Plant lice and mealy bugs. – Rep. State Entom. Illinois, 18: 55–108.
Harten, A. van & Ilharco, F. A., 1971: Notes on the aphid fauna of Angola, with the description of a new species of *Schizaphis* Börner (Homoptera, Aphidoidea). – Rev. Ciênc. Biol. ser. A, 3 (1970): 1–24.
– & – , 1972: Recent additions to the aphid fauna of Angola, including a new species of *Antalus* Adams (Homoptera, Aphidoidea). – Ibid., 4 (1971): 107–121.
— , 1976: A further contribution of the aphid fauna of Angola, including the description of a new genus and species (Homoptera, Aphidoidea). – Agronomia Lusitania, 37(1): 13–25.
Hartig, T., 1839: Jahresberichte über die Fortschritte der Forstwissenschaft und forstlichen Naturkunde im Jahre 1836/37 nebst Original-Abhandlungen aus dem

Gebiete dieser Wissenschaft. – Berlin. Abhandl., I: 640–646.
– , 1841: Versuch einer Eintheilung der Pflanzenläuse (Phytophthires Burm.) nach der Flügelbildung. – Germar's Z. f. Ent., 3: 359–376.
Hausmann, T., 1802: Beiträge zur Geschichte der Insekten. – Illigers Mag., 1: 426–445.
Heie, O. E., 1960–1970: A list of Danish aphids. Parts 1–9. – Ent. Meddr, 29 (1960): 193–211; 31 (1961): 77–96; 31 (1962): 205–224; 32 (1964a): 341–357; 35 (1967a): 125–141; 37 (1969a): 70–94, 373–385; 38 (1970): 137–164, 197–214.
– , 1964b: Aphids collected in Iceland in August, 1961. – Ent. Meddr, 32: 220–235.
– , 1965: A new species of *Macrosiphum* from *Chamaenerium* (Homoptera, Aphididae). – Ibid., 34: 31–42.
– , 1967b: Studies on fossil aphids (Homoptera: Aphidoidea), especially in the Copenhagen collection of fossils in Baltic amber. – Spolia zool. Mus. Haun., 26.
– , 1969b: *Schizoneura patchiae* Börner & Blunck, 1916, and *S. patchi* Meunier, 1917, a case of virtual homony and a proposed solution. – Bull. zool. Nom., 25: 222–223.
– , 1972a: Nogle for Danmark nye bladlusarter (Homoptera: Aphidoidea). – Flora og Fauna, 78: 93–96.
– , 1972b: Færøernes bladlus (Homoptera, Aphidoidea). (Aphids of the Faroes). – Ent. Meddr, 40: 145–150.
– , 1973a: Tilføjelser til listen over danske bladlus (Homoptera, Aphidoidea). (Additions to the list of Danish aphids). – Ibid., 41: 177–187.
– , 1973b: Bladlus. – Natur og Museum, Århus, 15: 1–22.
– , 1976a: Taxonomy and phylogeny of the fossil family Elektràphididae Steffan, 1968 (Homoptera: Aphidoidea). – Ent. scand., 7: 53–58.
– , 1976b: *Masonaphis lambersi* MacGill. og andre for Danmark nye bladlus (Homoptera, Aphidoidea). – Ent. Meddr, 44: 3–8.
– , 1980: Morphology and phylogeny of some Mesozoic aphids (Insecta, Hemiptera). – Ent. scand., 11: – (in press).
– & Heikinheimo, O., 1966: Aphids collected in Finland during the 12th N. J. F. Congress in 1963. – Ann. Ent. Fenn., 32: 113–127.
Heikinheimo, O., 1944: Für die finnische Fauna neue Blattläuse (Homoptera, Aphidoidea). – Ibid., 10: 1–7.
– , 1946: Eine für die finnische Fauna neue, an Gramineenwurzeln lebende Blattlaus, *Forda formicaria* v. Heyd. – Ibid., 11: 234–235.
– , 1959: Kasviviruksia siirrostavien lehtikirvojen esiintymisestä maassamme. (On the occurrence of virus vector aphids in Finland). – Publ. Finnish State Agr. Res. Board, 178: 20–40.
– , 1963: Für die finnische Fauna neue Blattläuse (Homoptera, Aphidoidea). II. – Ann. Ent. Fenn., 29: 184–190.
– , 1966a: Für die finnische Fauna neue Blattläuse (Homoptera, Aphidoidea). III. – Ibid., 32: 107.
– , 1966b: Aphids (Homoptera, Aphidoidea) caught in Norway SFi Aurland under an excursion of the 13th congress of Fennoscandian entomologists, in August 14–16, 1965. – Norsk ent. Tidsskr., 13: 387–392.

Hellén, W., 1942–1966: Verzeichnis der in Jahren 1936–1965 für die Fauna Finnlands neuhingekommenen Insektenarten. – Not. Ent., Helsingfors, 21 (1942: in den Jahren 1936–1940), 26 (1946: in den Jahren 1941–1945), 32 (1952: in den Jahren 1946–1950), 36 (1956: in den Jahren 1951–1955), 41 (1961: in den Jahren 1956–1960), 1966 (in den Jahren 1961–1965).

Henriksen, K. L., 1944: Fortegnelse over de danske Galler (Zoocecidier). – Spolia zool. Mus. haun., 6: 1–212.

Heyden, C. H. G. von, 1837: Entomologische Beiträge. Hemiptera, Aphid. – Mus. Senckenbergianum Abhdl., 2: 289–299.

Higuchi, H. & Miyazaki, M., 1969: A tentative catalogue of host plants of Aphidoidea in Japan. – Insecta Matsumurana, Suppl. 5: 1–66.

Hille Ris Lambers, D., 1938: Contributions to a monograph of the Aphididae of Europe I. – Temminckia, Leiden, 3: 1–44.

– , 1939a: Contributions to a monograph of the Aphididae of Europe II. – Ibid., 4: 1–134.

– , 1939b: On some Western European aphids. – Zool. Meded., 22: 79–119.

– , 1947a: Contributions to a monograph of the Aphididae of Europe III. – Temminckia, Leiden, 7: 179–320.

– , 1947b: On some mainly Western European aphids. – Zool. Meded., 28: 291–333.

– , 1949: Contributions to a monograph of the Aphididae of Europe IV. – Temminckia, Leiden, 8: 182–323.

– , 1950a: Hostplants and aphid classification. – 8th Internat. Congr. Ent. Stockholm 1948: 141–144.

– , 1950b: An apparently unrecorded mode of reproduction in Aphididae. – Ibid., 235.

– , 1951: On mounting aphids and other soft-skinned insects. – Ent. Ber. 298, XIII: 55–58.

– , 1952: The aphid fauna of Greenland. – Meddr Grønland, 136(1): 1–33.

– , 1953: Contributions to a monograph of the Aphididae of Europe V. – Temminckia, Leiden, 9: 1–176.

– , 1955: Hemiptera 2. Aphididae. Zoology Iceland, 3 (52a): 1–29.

– , 1960: Additions to the aphid fauna of Greenland. – Meddr Grønland, 159 (5): 1–18.

– , 1964: Higher categories of the Aphididae. – Abstracts of the papers presented at the seminar on the current status of research of aphids. – Univ. Calif. Berkeley: 2. (Duplicated).

– , 1968: A study of *Neuquenaphis* Blanchard, 1939, with descriptions of new species (Aphididae, Homoptera). – Tijdschr. Ent., 111 (7): 257–286.

– , 1970: A study of *Tetraneura* Hartig, 1841 (Homoptera, Aphididae), with descriptions of a new subgenus and new species. – Boll. Zool. agr. Bachic., ser. II, 9, 1968–69: 21–101.

– , 1972: Aphids: their life cycles and their role as virus vectors. – *In* J. A. de Bokx (Ed.): Viruses of potatoes and seed-potato production. Pudoc, Wageningen, 1972: 36–56.

– , 1980: Aphids as botanists? – Symp. Bot. Upsal., 22 (1979): 114–19.

Holman, J., 1965: Some unrecorded Middle European aphids. – Acta ent. Mus. Nat. Pragae, 11: 277–284.
– , 1971: Vorläufige Übersicht der Aphidenfauna des Gebirges Novohradské hory. – Sb. Jihocesk. Muzea v Ceských Budejovicích Prírodn. Vedy Supplementum 11: 11–20.
– & Szelegiewicz, H., 1971: Notes on *Aphis* species (Homoptera, Aphididae) from Mongolia and the U.S.S.R., with descriptions of four new species. – Acta ent. bohemoslov., 68: 397–415.
– & – , 1972: Weitere Blattläuse (Homoptera, Aphidodea) aus der Mongolei. – Fragm. faun., 18 (1): 1–22.
Holzner, G., 1874: Vorläufige Mittheilung über *Pemphigus poschingeri* n.sp. Tannenwurzel-Laus. – Stettin. ent. Ztg., 35: 221–222.
Horvath, G. de, 1896: Eine alte und drei neue Aphiden-Gattungen. – Wien. ent. Ztg, 15: 1–7.
Ilharco, F. A., 1960: O conhecimento dos afídeos em Portugal (Hemiptera, Aphidoidea). – Broteria, 29: 150–174.
– , 1961: On an aphid collection found in the Estacao Agronómica Nacional, including a new species, *Paraschizaphis rosazedovi* (Hemiptera, Aphidoidea). – Agros, 44: 71–77.
– , 1964: A study on the systematic position of the genus *Israelaphis* Essig, with descriptions of the alate forms and the first instar nymphs of *Israelaphis lambersi* Ilharco (Homoptera – Aphidoidea). – Agron. Lusitania, 26: 257–272.
– , 1967–1969: Algunas correccoes e edicoes à lista de afídeos de Portugal Continental. I–V. (Homoptera – Aphidoidea). – Ibid. 29 (1967): 117–139, 221–245; 30 (1968): 23–34; 31 (1969): 341–348.
– , 1969: Notes on the aphid fauna of Mozambique. Part I. – Rev. Cienc. Biol., 2: 1–9.
– , 1971: Notes on the aphid fauna of Mozambique. Part II. – Ibid., 4: 123–127.
– , 1973: Catálogo dos afídeos de Portugal Continental. – Estacao Agronómica National, Oiras, 135 pp.
Johnson, B., 1958: Factors effecting the locomotor and settling responses of alate aphids. – Anim. Behaviour, 6: 9–26.
Jørgensen, M., 1932: Aphidae. Zoology Faroes, XLV: 1–9. Copenhagen.
Kaltenbach, J. H., 1843: Monographie der Pflanzenläuse (Phytophthires). – Aachen. xliii + 223 pp.
Kanervo, V., 1962: Über einen Fall von anemochorer Ausbreitung der Blattläuse in Finnland im Sommer 1959. – Verh. XI. Intern. Kongr. Wien 1960, Bd. III (1962): 20–21.
Kennedy, J. S. & Stroyan, H. L. G., 1959: Biology of aphids. – A. Rev. Ent., 4: 139–160.
Kislow, C. J. & Edwards, L. J., 1972: Repellent odour in aphids. – Nature, 235: 108–109.
Klingauf, F., 1972a: Einfluss von Aminosäuren auf das Wirtswahl-verhalten von *Acyrthosiphon pisum* (Homoptera: Aphididae) unter besonderer Berücksichtigung ihres chemischen Aufbaus. – Entomologia exp. appl., 15: 274–286.
– , 1972b: Die Bedeutung von peripher vorliegenden Pflanzensubstanzen für die

Wirtswahl von phloemsaugenden Blattläusen (Aphididae). – Z. Pflanzenkrankh. u. Pflanzenschutz, 79: 471–477.
Koch, C. L., 1854–1857: Die Pflanzenläuse Aphiden, getreu nach dem Leben abgebildet und geschrieben. – Nürnberg. 336 pp., 54 pl.
Kunkel, H., 1969: Über das Verhalten der Aphiden und verwandter Honigtauerzeuger bei der Abgabe von Honigtau. – Apimondia. XXII. Intern. Bienenzäuchter Kongr. München 1969: 477–481 (Bukarest).
– , 1972: Die Kotabgabe bei Aphiden (Aphidina, Hemiptera). – Bonn. zool. Beitr., 23: 161–178.
– , 1973: Die Kotabgabe der Aphiden (Aphidina, Hemiptera) unter Einfluss von Ameisen. – Ibid., 24: 105–121.
Krzywiec, D., 1962: Morphology and biology of *Mimeuria ulmiphila* (del Guercio) (Homoptera, Aphidina). Part I. – Bull. Biol. Poznan, Ser. D, Livr. III: 63–97.
Lampel, G., 1970: Über einen Blattlaus-Massenbefall an Pappeln in Schweden. – Natur und Museum, Frankfurt/Main, 100: 17–21.
– , 1974: Für die Schweiz neue Blattlaus-Arten (Homoptera, Aphidina) 1. – Mitt. schweiz. ent. Ges., 47: 273–305.
Landin, B.-O., 1967: Insekter I. – Fältfauna. – Stockholm.
Leach, W. E., 1818: Notes on the insect *Aphis lanigera*. – Trans. hort. Soc. Lond., 3: 60.
Léclant, F., 1966–1968: Contribution à l'étude des Aphidoidea du Languedoc Meridional. I–III. – Ann. Soc. Hort. Hist. Natur. Hérault, 106 (1966): 119–134; 107 (1967): 38–45; 108 (1968): 138–143.
Lees, A. D., 1959: The role of photoperiod and temperature in the determination of parthenogenetic and sexual forms in the aphid *Megoura viciae* Buckton. – I. The influence of these factors on apterous virginoparae and their progeny. – J. Insect Physiol., 3: 92–117.
– , 1961: Clonal polymorphism in aphids. – *In:* J. S. Kennedy (ed.): Insect polymorphism. R. ent. Soc. Lond., 1961: 68–79.
– , 1966: The control of polymorphism in aphids. – Advances in Insect Physiology. London and New York. Academic Press. Vol. 3: 207–277.
– , 1967a: Direct and indirect effects of day length on the aphid *Megoura viciae* Buckton. – J. Insect Physiol., 13: 1781–1785.
– , 1967b: The production of the apterous and alate forms in the aphid *Megoura viciae* Buckton, with special reference to the role of crowding. – Ibid., 13: 289–318.
Lichtenstein, J., 1885: Les Pucerons. Monographie des Aphidiens (Aphididae Passerini, Phytophtires Burmeister). Première partie. – Genera. – Montpellier. 188 pp.
Lid, J., 1963: Norsk og svensk flora. – Oslo. 800 pp.
Lindblom, A., 1936: Skadedjur i Sverige år 1935. – Stat. växtskyddsanst. medd., 16.
– , 1938: Skadedjur i Sverige år 1936. – Ibid., 26.
– , 1941: Skadedjur i Sverige år 1937. – Ibid., 35.
Linnaniemi, W. M., 1915: 19. Berättelse över skadeinsekters uppträdande i Finland år 1913. – Lantbruksstyrelsens Meddelanden, 99: 1–68.

– , 1916: 20. Berättelse över skadeinsekters uppträdande i Finland år 1914. – Ibid., 111: 1–75.
– , 1921: 21.–22. Berättelse över skadedjurs uppträdande i Finland under åren 1915 och 1916. – Ibid., 131, xxvii + 227 pp.
– , 1935: Kertomus tuhoeläinten esiintymisestä suomessa. Vuosina 1917–1923. (Bericht über das Auftreten der Pflanzenschädlinge in Finnland in den Jahren 1917–1923). – Valtion Maatalouskoetoimennan Julkaisuja, Helsinki, 68 (Die staatliche landwirtschaftliche Versuchstätigkeit, Veröffentlichung 68): 1–159 (Aphididae: 117–130, 150, 154–158).
Linné, C. von, 1758: Systema Naturae. Editio decima, reformata 1: 1–824.
– , 1761: Fauna suecica sistens animalia Sueciae regni. Editio altera, auctior 1761.
– , 1767: Systema Naturae. Editio duodecima, reformata. 1(2): 533–1327.
Lundberg, J., 1963: Cecidiestudier. – Opusc. ent., 28.
Lundblad, O., 1927: Skadedjur i Sverige åren 1922–1926. – Medd. Centr. anst. förs.väs. jordbr.omr., Stockholm, 317.
– , 1928: Skadedjur i Sverige år 1927. – Ibid., 337.
– & Tullgren, A., 1923: Skadedjur i Sverige åren 1917–1921. – Ibid., 259.
Mackauer, M., 1965: Parasitological data as an aid in aphid classification. – Can. Ent., 97: 1016–1024.
Marchal, P., 1933: Les Aphides de l'orme et leurs migrations. – Ann. Epiphyt., 19: 207–329.
Matsumura, S., 1917: A list of the Aphididae of Japan, with description of new species and genera. – J. Coll. Agr. Tokoku Imp. Univ., 7: 351–414.
Meijere, J. C. H. de, 1912: Zur Kenntnis von *Hamamelistes betulae* Mordvilko. – Z. wiss. Insekt Biol., 8: 89–94.
Mier Durante, M. P., 1978: Estudio de la Afidofauna de la Provincia de Zamora. – Caja de Ahoros Provincial de Zamore. 226 pp.
Miles, P. W., 1965: Studies on the salivary physiology of plant bugs: The salivary secretions of aphids. – J. Insect Physiol., 11: 1261–1268.
– , 1968: Insect secretions in plants. – Ann. Rev. Phytopath., 6: 137–164.
Mittler, T. E., 1958: Studies on the feeding and nutrition of *Tuberolachnus salignus* (Gmelin). III. The nitrogen economy. – J. Exp. Biol. 35: 626–638.
Miyazaki, M., 1971: A revision of the tribe Macrosiphini of Japan (Homoptera: Aphididae, Aphidinae). – Insecta Matsumurana, 34(1): 1–247.
Moericke, V., 1941: Zur Lebensweise der Pfirsichblattlaus (*Myzodes persicae* Sulz.) auf der Kartoffel. – Bonn. 101 pp.
– , 1955: Über die Lebensgewohnheiten der geflügelten Blattläuse (Aphidina) unter besonderer Berücksichtigung des Verhaltens beim Landen. – Z. angen. Ent., 37: 29–91.
Monell, J., 1877: A new genus of Aphidae. – Can. Ent., 9: 102–103.
– , 1882: Note on Aphididae. – Ibid., 14: 13–16.
Mordvilko, A. K., 1895: K faune i anatomii sem. Aphididae Privislanskogo Kraja. – Vars. Univ. Izv., 1895 (107): 113–274.

– , 1901: Zur Biologie und Morphologie der Pflanzenläuse (Fam. Aphididae Pass.). – Horae Soc. Entomol. Ross., 33: 162–302, 303–1012.
– , 1908: Tableaux pour servir à la determination des groupes et des genres des Aphididae Pass. – Ann. d. Mus. Zool. de l'Acad. Imp. d. Sciences, St. Petersburg, 13: 353–384.
– , 1916: On some old and new aphids. – Russk. Entomol. obozrenne, 15 (1915): 75–79.
– , 1928a: *Geoica* Hart and its anholocyclic forms. – Comp. Rend. Acad. Sci. U.S.S.R. (A), 1928: 525–528.
– , 1928b: Nouvelle contribution à l'étude de l'anolocyclie chez les Aphidés. – *Forda formicaria* Heyden et sa forme anolocyclique. – Compt. Rend. Acad. Sci. Paris, 187: 1070–1072.
– , 1928c: Les pemphigiens des pistachiers et leur formes anolocycliques. – Bull. Soc. zool. Fr., 53: 358–366.
– , 1928d: Podotrjad Aphidoidea, tli ili ratsitelnye vsi. – *In:* I. N. Filipjev: Oprede litelj nasekomych. 943 pp.
– , 1931: Heteroecious and anolocyclic Anoeciinae. – Izv. A. N. S.S.S.R., 1931: 871–880.
– , 1935: Die Blattläuse mit unvollständigem Generationszyklus und ihre Entstehung. – Ergebn. Fortschr. Zool., 8: 36–328.
Müller, F. P., 1955: Blattläuse. – Neue Brehm-Bücherei, Wittenberg-Lutherstadt.
– , 1961: Ergänzungen zur Blattlausfauna von Mitteleuropa. – Mitt. dt. ent. Ges., 20: 69–70.
– , 1962: Hemiptera (Homoptera): Aphididae. – S. Afr. Anim. Life, 9: 319–323.
– , 1964: Merkmale der in Mitteleuropa an Gramineen lebenden Blattläuse (Homoptera: Aphididae). – Wiss. Z. Univ. Rostock, 13, Naturwiss. Reihe, Hft. 2/3: 269–278.
– , 1968: Weitere Ergänzungen zur Blattlausfauna von Mitteleuropa (Homoptera, Aphidina). – Faun. Abhandl., 2: 101–106.
– , 1969: Aphidina – Blattläuse, Aphiden. – *In:* E. Stresemann: Exkursionsfauna von Deutschland. Wirbellose II/2, Berlin, pp. 51–141.
Nasonov, N. V., 1894: Collection of Zoological Museum II. Warszawa, 1894.
Nault, L. R., 1973: Alarm pheromones help aphids escape predators. – Ohio Report, Wooster, Ohio, 58: 16–17.
– , Edwards, L. J. & Styer, W. E., 1973: Aphid alarm pheromones: Secretion and reception. – Environm. Entomol., 2: 101–105.
Nordman, A., 1959: Fynd av bladlusarten *Paracletus cimiciformis* Heyd. i övervintringsbon av *Tetramorium cespetum* L. (Hem.). – Notul. ent., 39.
Nuorteva, P., 1961: *Pachypappella populi* (L.) (Hom., Pemphigidae) as a pest of *Populus tremula* (L.) in Finland. – Ann. Ent. Fenn. 27: 123–126.
Oestlund, O. W., 1919: Contribution to knowledge of the tribes and higher groups of the Aphididae (Homoptera). – Rep. State Entomol. (Minnesota), 17: 46–72.
Ossiannilsson, F., 1941: Några för Sverige nya eller hos oss föga beaktade Hemiptera. – Opusc. ent., 6.

– , 1943: Hemipterologiska notiser. – Ibid., 8.
– , 1951: Hemiptera. – *In:* Brinck-Wingstrand: The mountain fauna of the Virihaure area in Swedish Lapland. II. Special account. – K. Fysiogr. Sällsk. Handl. N. F., 2.
– , 1959: Contributions to the knowledge of Swedish aphids. I–II. – Kungl. Lantbrukshögskolans Ann., 25: 1–46, 375–527.
– , 1961: Anmärkningar och tillägg till Sveriges hemipterfauna (Hemipterologiska notiser VIII). – Opusc. ent., 26.
– , 1962: Hemipterfynd i Norge 1960. – Norsk ent. Tidsskr., 12: 56–62.
– , 1964a: Contributions to the knowledge of Swedish aphids. III. List of food plants. – Kungl. Lantbrukshögskolans Ann., 30.
– , 1964b: Anmärkningar och tillägg till den svenska bladlusfaunan (Hemiptera, Aphidoidea). – Ent. Tidskr., 85.
– , 1969a: Notes on some Swedish aphids (Hem., Aphidoidea). – Opusc. ent., 34.
– , 1969b: Catalogus Insectorum Sueciae XVIII. Homoptera: Aphidoidea. – Ibid., 34: 35–72.
– , 1978: The Auchenorrhyncha (Homoptera) of Fennoscandia and Denmark. Part I: Introduction, infraorder Fulgoromorpha. – Fauna ent. scand., 7, part 1: 1–222.
Osten-Sacken, C. R., 1861: Über die Gallen und andere durch Insecten hervorgebrachte Pflanzendeformationen in Nord-America. – Stettin. ent. Ztg., 22: 405–423.
Paik, W. H., 1965: Aphids of Korea. – Seoul. 160 pp.
Palmer, M. A., 1952: Aphids of the Rocky Mountain region. – Thomas Say Foundation, Denver, vol. 5. 452 pp.
Parker, A. N. B., 1976: Unpublished thesis. Harpenden, England.
Passerini, G., 1856: Gli insetti autori delle galle del Terebinto e del Lentisco insieme ad alcune specie congeneri. – Giardini, G. Orticolt, 3: 258–265.
– , 1857: Gli Afidi. – Ibid., 7: 1–20.
– , 1860: Gli Afidi con un prospetto dei generi ed alcune specie nuove Italiane. – Parma. 46 pp.
– , 1861: Additamenta ad indicem Aphidinarum quas hucusque in Italia lectarum. – Atti Soc. Italiana Sci. Nat. Milano, 3: 398–401.
– , 1863: Aphididae Italicae hucusque observatae. – Arch. Zool. Anat. Fisiol. Modena, 2: 129–212.
Pettersson, J., 1970: An aphid sex attractant. – Ent. scand., 1: 63–73.
Prior, R. N. B. & Stroyan, H. L. G., 1960: On a new collection of aphids from Iceland. – Ent. Meddr, 29: 266–293.
Raychaudhuri, D. N., 1969: Taxonomy of the aphids of the eastern Himalayas. – U.S. PL-480 Project. Report 1968–69. Calcutta. 16 pp.
Remaudière, G., 1951: Contribution à l'étude des Aphidoidea de la faune francaise. Aphididae: Dactynotinae et Myzinae. – Rev. Pathol. Vég. Entomol. Agr. Fr., 30: 125–144.
– , 1952: Contribution à l'étude des Aphidoidea de la fauna francaise. Déscription de quelques Aphididae nouveaux et addition à la liste des Myzinae et Dactynotinae. –

Ibid., 31: 232–263.
– , 1954: Deuxième addition à la liste des Dactynotinae et Myzinae (Hom. Aphidoidea) de la faune francaise. – Ibid., 33: 232–240.
Reuter, O. M., 1883: (*Schizoneura ulmi* recorded from Finland). – Medd. F. Fl. Fenn. 9.
Richards, W. A., 1964: On the evolution, origin and dispersal of the Panaphidines. – Abstracts of the Papers Presented at the Seminar on the Current Status of Research of Aphids. Univ. Calif., Berkeley: 6 (duplicated paper).
Robinson, A. G., 1972: Annotated list of aphids (Homoptera: Aphididae) collected in Thailand, with description of a new genus and species. – Can. Ent., 104: 603–608.
Rostrup, S., 1897: Danske Zoocecidier. – Vidensk. Meddr dansk naturh.Foren. 1896: 1–64.
– , 1900: Vort Landbrugs Skadedyr. – Copenhagen. (1940: Fifth edition by Bovien, P. & Thomsen, M.).
Rondani, C., 1848: Osservazioni sopra parecchie specie di Esapoda Aficidi e sui loro nemici. – Nuove Ann. Sc. Nat. Bologna (Ser. 2), 8: 432–448.
Schouteden, H., 1906: Cataloque des Aphides de Belgique. – Mém. Soc. ent. Belg., 12: 189–246.
Schöyen, T. H., 1914–1921: Beretning om skadeinsekter og plantesygdomme i land- og havebruget. – Årsberetn. off. Foranst. Landbr. Frem. 1913–1920.
– , 1922–1941: Melding on skadeinsekter i jord- og hagebruk. – Ibid., 1920–1939.
Schöyen, W. M., 1893–1912: Beretning om Skadeinsekter og Plantesygdomme i Land- og Havebruget. – Ibid., 1892–1912.
Schumacher, F., 1923: *Paracletus cimiciformis* Heyd., die *Tetramorium*-Wurzellaus. – Dt. ent. Z., 4: 401–410.
Schwartz, H., 1932: Der Chromosomenzyklus von *Tetraneura ulmi* De Geer. – Z. Zellforsch., 15: 645–687.
Semal, J., 1956: Catalogue des Aphididae de Belgique. – Bull. Annls Soc. r. ent. Belg., 92: 79–94.
Shaposhnikov, G. K., 1964: Suborder Aphidinea – Plant lice. – *In:* G. Ya. Bei-Bienko et al.: Keys to the insects of the European U.S.S.R. (in Russian, English translation: Jerusalem 1967: 616–799).
Shimer, H., 1867: On a new genus of Aphidae. – Trans. Am. ent. Soc., 1: 283–285.
Siebke, H., 1874: Enumeratio Insectorum Norvegiorum. Fasc. I. – Christiania.
Simmonds, S. P., 1956: *Eriosoma patchae* on Cinerarias. – Plant Pathol., 5: 76.
Smith, C. F., 1972: Bibliography of the Aphididae of the world. – North Carolina Agr. Exp. Sta. Tech. Bull., 216: 1–717.
– , & Cermeli, M. M., 1979: An annotated list of Aphididae (Homoptera) of the Carribbean Islands and South and Central America. – Ibid., 259: 1–131.
– , & Parron, C. S., 1978: An annotated list of Aphididae (Homoptera) of North America. – Ibid., 255: i–viii, 1–428.
Steffan, A. W., 1968: Elektraphididae, Aphidinorum nova familia e sucino baltico (Insecta: Homoptera: Phylloxeroidea). – Zool. Jb. Syst. Mainz, 95: 1–15.
Stroyan, H. L. G., 1950: Recent additions to the British aphid fauna. Part I: *Dactynotus*

Rafinesque to *Rhopalosiphum* Koch, C. L. – Trans. R. ent. Soc. Lond., 101: 89–123.
– , 1955: Recent additions to the British aphid fauna. Part. II. – Ibid., 106: 283–340.
– , 1957: Further additions to the British aphid fauna. – Ibid., 109: 311–360.
– , 1964a: Notes on hitherto unrecorded or overlooked British aphid species. – Ibid., 116: 29–72.
– , 1964b: Aphidoidea (excluding Adelgidae & Phylloxeridae). – *In:* G. S. Kloet & W. D. Hincks: A check list of British insects. 2nd edition. Roy. ent. Soc. Lond., 1: 67–86.
– , 1964c: Notes on some British species of *Pemphigus* Hartig (Homoptera: Aphidoidea) forming galls on poplar, with the description of a new species. – Proc. R. ent. Soc. Lond. (B), 33: 92–100.
– , 1966: Notes on aphid species new to the British fauna. – Ibid., 35: 111–118.
– , 1972: Additions and amendments to the check list of British aphids (Homoptera: Aphidoidea). – Trans. R. ent. Soc. Lond., 124: 37–79.
– , 1975: The life cycle and generic position of *Aphis tremulae* L., 1761 (Aphidoidea: Pemphiginae), with a description of the viviparous morphs and a discussion of spruce root aphids in the British Isles. – Biol. J. Linn. Soc., 7: 45–72.
– , 1977: Homoptera: Aphidoidea. Chaitophoridae and Callaphididae. – Handbk Ident. Br. Insects. II, 4(a), 130 pp.
Suomalainen, E., 1935: Aphidina. – *In:* Enumeratio Insectorum Fenniae. III. Hemiptera: 12–13. Helsingfors.
Sutherland, O. R. W., 1969: The role of crowding in the production of winged forms by two strains of the pea aphid, *Acyrthosiphon pisum.* – J. Insect Physiol., 15: 1385–1410.
Sylvén, E., 1950: Anmärkningsvärda skadedjur i Skåne 1950. – Växtskyddsnotiser, 14.
– , 1929: Kulturväxterna och djurvärlden. Svenska jordbrukets bok. – Stockholm.
Szelegiewicz, H., 1968a: Notes on some aphids from Vietnam, with description of a new species (Homoptera, Aphidodea). – Annls zool., 25: 459–471.
– , 1968b: Katalog Fauny Polski. Mszyce – Aphidoidea. – Polska Akad··Nauk. 316 pp.
– , 1968c: Faunistische Übersicht der Aphidoidea (Homoptera) von Ungarn. – Fragm. faun., 15: 57–98.
– , 1969: Mszyce – szkodniki roślin wektory chorób wirusowych i producenci spadzi. – Bibliografia. – Warszawa. 250 pp.
– , 1977: Levéltetvek I. – Aphidinea I. Magyarország Állatvilága – Fauna Hungariae. 128. XVII. Kötet. Heteroptera, Homoptera. 18. Füzet, 4. alrend., Budapest. 175 pp.
– , 1978a: Klucze do oznaczania owadon polski. Cześć XVII. Pluskwiaki równoskrzydle – Homoptera. Zeszyt 5a. Mszyce – Aphidodea – Lachnidae. – Polskie Towarzystwo Entomogiczne nr. 101. Warszawa. 107 pp.
– , 1978b: Przeglad systematyczny mszyc polski (A check list of the aphids of Poland). Zeszyty problemowe postepów nauk rolniczych. Zeszyt 208. – Polska Akademia Nauk Wydzial Nauk Rolniczych i Leśnych, Warszawa. 40 pp.
Tao, C. Chia-chu, 1961–70: Aphid fauna of China. – Science Yearbook of Taiwan Museum IV (1961): 35–44; V (1962): 33–82; VI (1963): 104–147; VII (1964): 36–74; VIII (1965): 1–28; IX (1966): 1–28; X (1967): 1–28; XI (1968): 1–56; XII (1969): 40–99; XIII (1970): 1–44.

Tambs-Lyche, H., 1968: Studies on Norwegian aphids (Hom., Aphidoidea). I. The subfamily Dactynotinae Börner. – Norsk ent. Tidsskr., 15: 1–17.
– , 1970: Studies on Norwegian aphids (Hom., Aphidoidea). II. The subfamily Myzinae (Mordvilko) Börner. – Ibid., 17: 1–16.
Theobald, F. V., 1914: Two new myrmecophilous aphids from Algeria. – Entomologist, 47: 28–31.
– , 1926–1929: The Plant Lice or Aphididae of Great Britain. I–III. London.
Thomsen, M. & Bovien, P., 1933: Haveplanternes Skadedyr. Copenhagen.
Thomson, C. G., 1862: Skandinaviens Insekter, en handbok i Entomologi. – Lund.
Thuneberg, E., 1960–1966: Beiträge zur Kenntnis der finnischen Blatt- und Schildläuse (Hom., Aphidoidea et Coccoidea) sowie deren Parasiten. I–IV. – Annls Ent. Fenn., 26 (1960): 97–99; 28 (1962): 40–43; 29 (1963): 130–134; 32 (1966): 153–158.
Tuatay, N. & Remaudière, G., 1964: Première contribution au catalogue des Aphididae (Homoptera) de la Turquie. – Rev. Pathol. Vég. Entomol. Agr. Fr., 43: 243–278.
Tullgren, A., 1909: Aphidologische Studien. – Ark. Zool., 5 (14): 1–190.
– , 1925: Aphidologische Studien. II. – Medd. Centr. Anst. Försöksv. Jordbr., 280.
– & Wahlgren, E., 1920–1922: Svenska insekter. – Stockholm.
Wahlgren, E., 1915, 1917: Det öländska alvarets djurvärld I–II. – Ark. Zool., 9, 11.
– , 1935–1961: Cecidiologiska anteckningar III & VI–XIII. – Ent. Tidskr., 56 (1935), 72 (1951), 74 (1953), 75 (1954), 77 (1957), 78 (1957), 79 (1959), 81 (1960), 82 (1961).
– , 1938: Svenska bladlöss (Aphidina). – Ibid., 59: 166–187.
– , 1939: Revision von Zetterstedt's lappländischen Aphidina. – Opusc. ent., 4: 1–9.
– , 1940: Till kännedomen om Skånes bladlusfauna. – Ibid., 5: 25–32.
– , 1954 & 1956: Die von Blattläusen erzeugten Pflanzengallen. I–II. Aphidocecidien in Ross-Hedicke: »Die Pflanzengallen Mittel- und Nordeuropas«. – Ibid., 19: 103–149; 21: 31–55.
– , 1955 & 1957: Aphidologiska notiser. – Ibid., 20: 1–9, 22: 126–135.
– , 1956b: Über *Byrsocrypta personata* Börn. – Ibid., 21: 147–148.
Walsh, B. D., 1867: On the insects, coleopterous, hymenopterous and dipterous, inhabiting the galls of certain species of willow. Part 2nd and last. – Proc. ent. Soc. Philad., 6: 223–288.
Walker, F., 1848: Remarks on the migrations of Aphides. – Ann. Mag. nat. Hist. (2), 1: 372–373.
Wearing, C. H., 1972: Selection of brussels sprouts of different water status by apterous and alate *Myzus persicae* and *Brevicoryne brassicae* in relation to the age of leaves. – Entomologia exp. appl., 15: 139–154.
Weber, H., 1928: Skelett, Muskulatur und Darm der Schwarzen Blattlaus *Aphis fabae* Scop. – Zoologica, 76: 1–120.
– , 1930: Biologie der Hemipteren. – Berlin, vii + 543 pp.
Westwood, J. O., 1840: An introduction to the modern classification of insects; founded on the natural habits and corresponding organisation of the different families. Vol. II. Aphidae, genera p. 118, species pp. 437–442.

– , 1849: Wingless subterranean plant lice. – Gard. Chron., 1849 (27): 420.
– , 1879: The fimbriated scale. – Ibid., 2 (12): 796.
Zetterstedt, J. W., 1828: Ordo III. Hemiptera. Fauna Insectorum Lapponica 1: i–xx, 1–563.
– , 1840: Aphidiae Latr. Insecta Lapponica descripta. Voss, Lipsiae 1840: i–vi, 1–1140.
Ziegler, H. & Mittler, T. E., 1959: Über den Zuckergehalt der Siebröhren- bzw. Siebzellensäfte von *Heracleum Mantegazzianum* und *Picea abies* (L.) Karst. – Z. f. Naturforsch., 14b: 278–281.
Zimmerman, E. C., 1948: Homoptera: Sternorrhyncha. – Insects of Hawaii, 5: i–vii, 1–464.
Zwölfer, H., 1957–1958: Zur Systematik, Biologie und Ökologie unterirdisch lebender Aphiden (Homoptera, Aphidoidea). I–IV. – Z. angew. Ent., 40 (1957): 182–221, 528–575; 42 (1958): 129–172; 43 (1958): 1–52.

Index

Synonyms are given in italics. The number in bold refers to the main treatment of the taxon.

Author's address:

Ole E. Heie
Erantisvej 32
DK-7800 Skive
Denmark

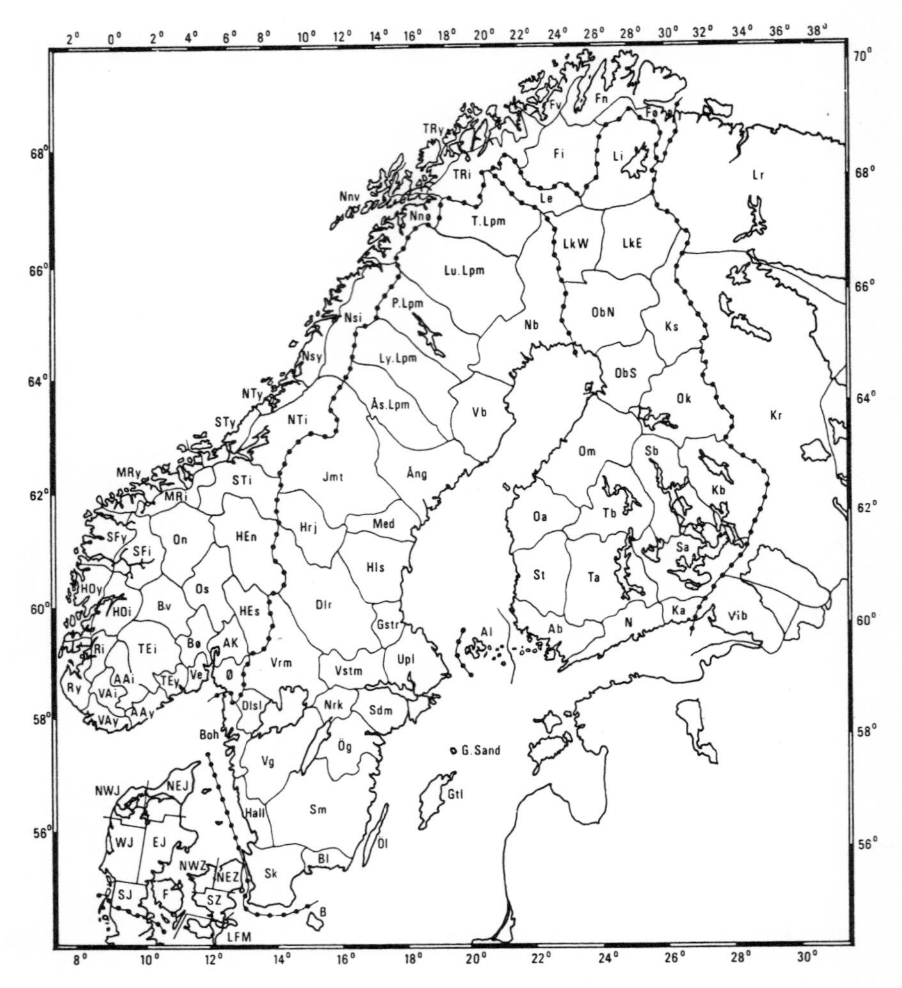

List of abbreviations for the provinces used throughout the text, on the map and in the following tables.

DENMARK

SJ	South Jutland	LFM	Lolland, Falster, Møn
EJ	East Jutland	SZ	South Zealand
WJ	West Jutland	NWZ	North West Zealand
NWJ	North West Jutland	NEZ	North East Zealand
NEJ	North East Jutland	B	Bornholm
F	Funen		

SWEDEN

Sk.	Skåne	Vrm.	Värmland
Bl.	Blekinge	Dlr.	Dalarna
Hall.	Halland	Gstr.	Gästrikland
Sm.	Småland	Hls.	Hälsingland
Öl.	Öland	Med.	Medelpad
Gtl.	Gotland	Hrj.	Härjedalen
G. Sand.	Gotska Sandön	Jmt.	Jämtland
Ög.	Östergötland	Ång.	Ångermanland
Vg.	Västergötland	Vb.	Västerbotten
Boh.	Bohuslän	Nb.	Norrbotten
Dlsl.	Dalsland	Ås. Lpm.	Åsele Lappmark
Nrk.	Närke	Ly. Lpm.	Lycksele Lappmark
Sdm.	Södermanland	P. Lpm.	Pite Lappmark
Upl.	Uppland	Lu. Lpm.	Lule Lappmark
Vstm.	Västmanland	T. Lpm.	Torne Lappmark

NORWAY

Ø	Østfold	HO	Hordaland
AK	Akershus	SF	Sogn og Fjordane
HE	Hedmark	MR	Møre og Romsdal
O	Opland	ST	Sør-Trøndelag
B	Buskerud	NT	Nord-Trøndelag
VE	Vestfold	Ns	southern Nordland
TE	Telemark	Nn	northern Nordland
AA	Aust-Agder	TR	Troms
VA	Vest-Agder	F	Finnmark
R	Rogaland		

n northern s southern ø eastern v western y outer i inner

FINLAND

Al	Alandia	Kb	Karelia borealis
Ab	Regio aboensis	Om	Ostrobottnia media
N	Nylandia	Ok	Ostrobottnia kajanensis
Ka	Karelia australis	ObS	Ostrobottnia borealis, S part
St	Satakunta	ObN	Ostrobottnia borealis, N part
Ta	Tavastia australis	Ks	Kuusamo
Sa	Savonia australis	LkW	Lapponia kemensis, W part
Oa	Ostrobottnia australis	LkE	Lapponia kemensis, E part
Tb	Tavastia borealis	Li	Lapponia inarensis
Sb	Savonia borealis	Le	Lapponia enontekiensis

USSR

Vib Regio Viburgensis Kr Karelia rossica Lr Lapponia rossica